Joe Pelanconi gives resonance to the story of an 1870s quicksilver mining boom in the mountains above the wine valleys of Sonoma County. This intriguing history sifts the ore of disparate and inaccessible source material, extracting a shining flask of Cinnabar narrative. His mini-biographies portray eccentric and curious miners, stage drivers and camp followers who clambered up the slopes of the Mayacamas as the price of quicksilver spiked and plummeted.
—Bo Simons, Sonoma County librarian, manager Healdsburg Regional Library

This engaging and well-researched California history book brings the Wild West mining village of Pine Flat to life with compelling stories of the colorful characters, hard labor and riches that accompanied the quicksilver boom of the 1870s. A must-read for any student of California history!
—Holly Hoods, curator, Healdsburg Museum and Historical Society

QUICKSILVER MINING

IN

SONOMA COUNTY

Pine Flat Prospect Fever

JOE PELANCONI

Charleston · London

THE
History
PRESS

Published by The History Press
Charleston, SC 29403
www.historypress.net

First published 2014

Manufactured in the United States

ISBN 978.1.62619.472.4

Library of Congress CIP data applied for.

For Mickey Bitsko, a dear friend, always there when needed. Bitsko lives in a fantasy world and does not read nonfiction, so if no one tells him about this, he'll never know.

CONTENTS

ACKNOWLEDGEMENTS

Much of the research for this work was originally done years ago as part of a master's degree thesis in U.S. history at CSU–Chico. I am indebted to long-retired CSU professors of history Dr. Lois Christensen, Dr. Clarence McIntosh and W.H. Hutchinson for their patience and diligence in assisting a somewhat reluctant student with learning the rigors of historical research. Also at that time, longtime Healdsburg historian Edwin Langhart provided valuable assistance.

More recently, Holly Hoods, curator of the Healdsburg Museum, was invaluable in making museum resources available and encouraging the research process. Old-timers like Harry Bosworth, Dint Rose and Dick Dilworth, with their stories and artifacts, provided inspiration. Harry also provided access to the subterranean depths of his general store, a place that inspires historical thought and is a repository for old photos, artifacts and cobwebs. Geyserville High School teacher and techie Rick Klug was helpful in providing technical support. Ann Howard provided skilled editorial assistance.

My charming wife, Sue, and two neurotic dogs did not impede my efforts. It should be further noted that in 1969, Sue, armed with whiteout, carbon paper and an archaic Underwood, typed my original thesis manuscript. For that alone, she should be allowed to buy new shoes with whatever profits result from this work.

Chapter 1

CHARACTERS OF
THE TIME

Nothing in history happens in a vacuum. One must understand the connection between a historical event and the time period. More to the point, one must understand how the main characters in an event were influenced by the times in which they lived. The historical record of the quicksilver rush of Sonoma County, California, from 1873 to 1875 was not difficult to record. Numerous newspaper accounts and official county records detailed the civil discourse, formal transactions and special events that took place in the Mayacamas Mountains. More challenging was the task of attempting to understand the context in which the events occurred and how that time frame flavored the historical record and shaped and motivated the participants.

The America of the 1870s was fascinated with the notion of a Manifest Destiny, led by the Anglo-Saxon will to subdue and fill the vast area from the Atlantic to the Pacific. It was an ethnically specific movement that spurred a frontier spirit of rugged individualism. "Go West, young man," was a battle cry to thousands who believed it was God's will that they head west, better themselves, make their own rules and decimate whoever or whatever was in their way.

During the decade of the 1870s, Sonoma County was a rural, agrarian enclave made up of farmers, merchants and fortune seekers, few of whom were native to the area but nearly all of whom to a degree had embraced a frontier spirit. Wappo and Pomo Native Americans had been rousted from their ancestral homeland. Early settlements by the Spanish, Russians and

Mexicans had been displaced by the influx of Americans, most with Anglo-Saxon lineage. Racism and nativism were endemic to 1870s America. The Declaration of Independence stated that "all men are created equal," but most white Americans assigned a less than human status to Chinese, Africans, Mexicans and Native Americans. Viewing those who were ethnically different as lower life forms justified prejudice and provided a convenient scapegoat for whatever troubles the Anglos encountered and, for fledgling entrepreneurs, a convenient source of cheap labor.

Thousands of single men came to California in search of wealth after the discovery of gold in 1848 and later adapted to the agrarian lifestyle when fortunes in mining were not forthcoming. Others came as families seeking free, fertile land, having made the sacrifices of a long and treacherous trek by wagon train from the Midwest. Some had fought in the Mexican-American War 1846–48, leading to California's statehood in 1850. Some had fought Indians on the way west. Others were Civil War veterans, both Union and Confederate, who had experienced the horrors of that conflict. Very few, if any, were Sonoma County native sons.

These men (and, to a lesser degree, women) were adventurous risk takers, seeking a new life in uncharted territory. They were optimistic, believing in hard work and instant wealth. They were men of action, adventurous and not well educated. They saw natural resources as unlimited and displayed few environmental concerns. Everything was new, and nothing seemed impossible. They were dreamers, possessing limited historical context in which to understand their impulses.

Although affected by the country's recession of the 1870s and still somewhat divided by the political climate that had led to the Civil War, they viewed national issues from a distance. They were comfortable making local decisions on the spot. The justice system was a work in progress with a lingering vigilante spirit. Local districts for schools, transportation and other public services were being created from scratch. Settlers were beginning to establish churches, schools and lodges. They were embracing the temperance spirit. The transcontinental railroad had been completed in 1872 and a rail line completed from San Francisco to Cloverdale.

Even though Sonoma County's populace was settling into a stable lifestyle, given their background, some of them remained only a wild promise from embracing any possibility of seeking an instant fortune. Overly optimistic journalists of the time did little to temper blind enthusiasm. Consequently, when one looks at the central figures in the quicksilver rush of 1873–75, one finds colorful characters with eventful personal histories ready to succumb to the lingering frontier spirit that brought them to California in the first place.

Newspaper accounts of the quicksilver rush, flavored with the optimistic spirit of the time, provided somewhat accurate accounts of the mining efforts and the building of Pine Flat. However, the journalists had no way of knowing, much less reporting, how history might view these colorful characters of the 1870s. The central players in the mining rush were ambivalent to California's history and open to whatever perceived opportunity lay ahead. They were the most courageous and visionary risk takers out of a generation of westerners all of whom harbored at least a part of the frontier spirit. Some would call them eccentric, others would say they were heroic. Often they were both. However defined, they remain intriguing and unique products of their time whom the noted "frontier thesis" historian Frederick Jackson Turner might have called "builders of the American character."

In an attempt to inject some historical context (and perhaps a bit of human interest) into the narrative of this story, I have chosen to highlight a few of the better-known participants with mini-biographies throughout the text. Since these biographies only reflect a few who were noteworthy enough to have been written about, they are not meant to amplify their actual significance in the rush itself. The rush undoubtedly attracted a number of other folks who made significant contributions but whose actions no one chose to record. They are simply meant to provide a glimpse into the life experiences of a few who were typical of the time and how their life experiences may have influenced their actions. They might also allow us to get a clearer view of what it actually might have been like to experience life in and about Pine Flat and the Cinnabar Mining District during the quicksilver rush of 1873–75.

A Character of the Time

ARCHIBALD CAMPBELL GODWIN

---◆►◄◆◄---

Archibald Campbell Godwin was born in Nansemond County, Virginia, in 1831, the son of Jonathan Lewis and Julia Campbell Godwin. As a one-year-old, he moved to Portsmouth, Virginia, where he was raised by his grandmother. As a strapping six-foot, six-inch nineteen-year-old, Godwin left Virginia in 1850 for the gold fields of California. Along with hundreds of other adventurous hopefuls, he set out to strike it rich in the newly formed state. Unlike most of the others, Godwin attained his goal. He found gold and by the early 1850s was a successful entrepreneur in gold, real estate, timber and cattle.

Once the gold rush subsided, the record of Godwin's immediate adventures is sketchy. Since he later became a noted brigadier general in the Confederate army, much was written of his military service. Many of those writers tried to reconstruct his pre-military exploits. There were accounts that he owned property on Vancouver Island, British Columbia; fought and killed Indians in Oregon; and was one vote from being a Democratic California governor nominee. While Godwin dealt in real estate, encountered hostile Indians and possessed strong political beliefs, much of this simply was not true. Nonetheless, firsthand accounts of those who encountered Godwin described him as a strong-willed, courageous and imposing figure.

Sometime in the early 1850s, Godwin landed in Northern California. By 1854, he had a home and mercantile business and raised livestock west of the Russian River in the Sonoma County stage stop called Godwin's Place, later named Clarville and eventually Geyserville. He was the first merchant in the community and quickly gained ownership of 640 acres in the nearby hills that included an area of significant geothermal activity known as the Geysers. His mercantile and livestock business did well, and visitors who did business with Godwin spoke of his "winning manners and personal magnetism."

As noted, the Geysers was part of Godwin's land acquisition. It had been discovered in 1847 by William Elliott and his son, who were pursuing a wounded bear. Sometime before 1854, Godwin explored the Geysers area and immediately saw its commercial value. He built a sawmill at the Geysers and used the lumber to construct a hotel

A.C. Godwin in his Confederate army uniform. *Courtesy Library of Congress.*

to accommodate curious excursionists. M. Levy, who already had a home in the area, ran Godwin's hotel and resort. The Geysers area was touted as a natural wonder rivaling Yosemite. It was said that Godwin lobbied to have the Godwin's Place stage stop name changed to Geyserville, partly to promote his hotel and resort enterprise.

With about 150 Native Americans living on the banks of the Russian River, Godwin was appointed the government Indian agent for the area. Government agents were charged with protecting the interests of white settlers and engaging the Indians in meaningful work. Godwin was mentioned in an 1855 government report to have been the "special agent on the Russian River" who reported that "whites threatened to drive the Indians out of the valley should they continue their refusal to work." C.F. Winslow, a visiting journalist, wrote of the Indians, "The disposition to deception and murder seems to be almost a universal quality of these untamable tribes." He further noted, "They are now, however, under Mr. Godwin's complete control." There is no evidence that Godwin disagreed with Winslow's pronouncements, and given his position, he was witness to and most likely a participant in the genocide of local native peoples and their brutal relocation to government reservations.

In 1855, a geologist desiring to visit the thermal area stopped to visit Godwin and secure supplies. Godwin led him on the fourteen-mile trek from his home to the thermal area. His report noted that Godwin was an "experienced hunter and marksman and is as familiar with the trails as the Indians or wild beasts themselves." He was so impressed with the "large Virginian hospitality" and "generous companion" that he insisted on naming the most conspicuous landmark Godwin's Peak, "by which it must be called by all travelers forever hereafter." That name stuck for several years, but after Godwin left the area, it became known as Geyser Peak.

Godwin noticed rich cinnabar ore deposits in the Geysers area and in 1859 established a mining district and staked a number of claims. Other men also attempted to stake claims. Maps were confusing, boundaries elusive and formal filings erratic, and Godwin ended up in a court hassle with John Baker over mining rights. Unwilling to pay legal fees to a lawyer, Godwin studied law, passed the legal licensing exam in 1860 and represented himself in the county courts in Santa Rosa. He won his case and hoped to begin profitable mining. However, with the low price of quicksilver, scarcity of labor and the lack of skill in reduction, Godwin was not able to show a profit. He finally chose to consolidate his claims, and some work was done, but with mounting losses, he put a stop to all work on the mines.

Godwin somehow found time to become involved in local politics. In 1855, he was elected a justice of the peace, and records indicate that he performed marriages as well as other duties. He was one of the prime movers in the building of a Sonoma County road to the Geysers. In a series of 1856 meetings in Santa Rosa, Godwin was selected to be on the board of directors of the Geyser Road committee that was created to select a route and raise funds. There were numerous references to his familiarity with the area. Of course, one should note that his keen interest may have been expected, since the proposed road would go through Godwin's Place up to the Geysers resort that he owned.

In 1861, Godwin's native state of Virginia seceded from the Union. In February of that year, a Godwin advertisement in the Petaluma Journal sought a renter for his geyser interests. He abruptly turned over his California holdings to friends and headed east. With the onset of the Civil War, Godwin joined the Confederate army and was commissioned a major. He was assigned as assistant provost marshal at Libby Prison in Richmond, Virginia. He was very successful in this position, and North Carolina authorities asked him to establish a prison camp in Salisbury, North Carolina, for two thousand Union prisoners of war in that state. He remained in that role until he was promoted to colonel and organized the Fifty-seventh North Carolina Regiment. He fought in battles at Fredericksburg, Chancellorsville and later

Gettysburg. He was captured and taken prisoner in 1863 and later released in exchange for a Union officer. He then returned to the Army of Northern Virginia and was promoted to brigadier general. While leading his troops against Union forces at Winchester, he was shot and killed on September 19, 1864. He was thirty-three years old.

Confederate war chronicles describe Godwin as a gallant and determined leader who never dreamed of surrender, even when confronted with overwhelming odds. On the day Godwin was captured, a member of his staff said the general heard a rumor of surrender and "immediately called for the man who made the declaration, and threatened to blow his brains out." Another footnote to his military service was his reputation for cruelty toward Union captives during his time heading the prisoner of war camp. After the war, there was Union talk of trying Godwin for war crimes until it was discovered he was already dead.

Godwin died a young man but with a list of accomplishments that were astounding. He did not spend much more than a half dozen years in Sonoma County, but he established the groundwork for much of what happened when the price of quicksilver skyrocketed in the 1870s. He was a leader in the removal of Native Americans from the area. With his development of the Geysers, he opened the area to better roads and hordes of visitors. His recognition of cinnabar deposits, staking of claims and initial mining efforts paved the way for the boom that was about to happen. Northern Sonoma County owed its inception and growth to his efforts. Godwin set the stage.

This hard-charging young man was a true character of the time. He left his Virginia home as a penniless teenager with little formal education. It is believed he never married. His documented treatment of Native Americans and prisoners of war suggests he was not a man of compassion—and compassion may not have been an attribute that would have served him well. Like thousands of others, he came west in search of wealth and adventure. His physical presence, native intelligence, magnetic personality and boundless energy trumped whatever deficiencies he harbored. In many ways, Godwin epitomized the most successful of the characters who built the West. A charismatic, ruthless and determined risk taker, he thought all things were possible—and for Archibald Campbell Godwin, most things were.[1]

Chapter 2

EARLY DISCOVERY AND
PROSPECTING FEVER

There are few students of western American history who have not studied in some detail the California gold rush and the Comstock silver rush of western Nevada. The economic importance, political significance and romantic lure of those two mining booms explain why they have attracted an overwhelming amount of historical attention. However, there were other mining episodes in western history that have been obscured and lost in the shadow of gold and silver. This is not meant to minimize the significance of gold and silver rushes or to exaggerate the relative importance of smaller, unstudied mining flurries but merely to suggest that gold and silver historians do not have a monopoly on mining excitement. In at least one localized and isolated case, quicksilver created nearly as much melodrama as was ever witnessed in the "days of '49."

Cinnabar ore, from which quicksilver or mercury is produced, is found only in areas of recent volcanic action and is therefore relatively scarce. California was known to possess rich deposits of cinnabar long before its admission to the Union in 1850. The Mexican government had discovered the red ore as early as 1824 and had mined significant quantities by 1845 at New Almaden in Santa Clara County. The Mexicans had been well aware of the fact that quicksilver had the unique property of amalgamating with gold or silver but did not compound with base elements of either. The amalgam of gold or silver with mercury then could be reduced by vaporizing the mercury in a retort. Because of this fact, the New Almaden deposits were developed soon after gold was discovered in California in 1848. By the mid-

1850s, New Almaden was the chief producer of quicksilver in the United States.[2] However, rich deposits of cinnabar were soon to be discovered in another section of California.

The Mayacamas District is a mountainous area about twenty miles in length located within the boundaries of three Northern California counties: Sonoma, Napa and Lake. The rugged and wooded area, situated about sixty-five air miles north of San Francisco, seemed quite safe from human intrusion when California became a state in 1850. However, there was one phenomenon of nature in the Sonoma County section of the district that soon led the curious into the brushy hills. While well known to local Native Americans, the steam-spewing geysers were first discovered by a white man in 1847. William Elliott called them "geysers," and the name stuck, even though the correct scientific name is "fumaroles." Others soon ventured into the area to get a firsthand view of steam violently belching from the earth. Most visitors waxed poetic, one calling the "Big Geysers the chemical laboratory of the Almighty." A few were less impressed, one writing, "They are certainly a curiosity, a marvel; but there is no element of beauty…like a three-legged calf, once seeing is satisfactory for a life-time."[3] Nonetheless, the Geysers were soon to rival Yosemite as California's greatest natural wonder and a significant tourist attraction.

One of the earliest witnesses was A.C. Godwin, who had been the first merchant in the nearby valley community called Godwin's Place. Godwin soon noticed abundant deposits of cinnabar on the ridge south of the

A postcard with a promotional photo of the Geysers. *Courtesy Healdsburg Museum.*

"The ✱ Geysers"

THE FAVORITE RESORT for tourists, sight-seers, health and pleasure seekers. A greater variety of mineral waters than in any other one place in the world. Varying in temperature from the cool mountain spring to 212° F. The only natural mineral, steam and Hammam baths. THEY PRODUCE A PERFECT COMPLEXION. A tepid swimming lake of mineral waters. The hotel and grounds embowered in fine specimens of California trees. Altitude 2000 feet. Atmosphere pure and free from fogs and wind. Not sufficiently inland to reach the heated belt. Fine fishing along the shady Pluton River which flows for two miles through the grounds. Every one desiring comfort, luxury, a good time and a view of the grandest scenery in the world should visit "THE GEYSERS." Only 96 miles from San Francisco, through the most picturesque part of California, including the Sonoma and Napa Valleys.

The grandest, CHEAPEST and most beneficial trip in the world: Round trip ticket and ONE WEEK at "THE GEYSERS," only $23.50. A complete change of air and a haven of rest, recreation and renovation for persons on the coast and in the interior valleys.

TELEPHONE CONNECTIONS with Western Union at Cloverdale and Calistoga.

ROUND TRIP TICKETS, unlimited, $8.50. TERMS: $3.00 per day, or $15.00 per week.

ALL BATHS FREE to guests furnishing their own bathing suits.

John B. Treadwell, Manager,

Geyser Springs, Sonoma Co., Cal.

A postcard with promotional information on the Geysers Resort. *Courtesy Healdsburg Museum.*

Geysers. Although he knew very little about quicksilver, Godwin nevertheless organized a mining district in 1859. The district included the western section of the Mayacamas Mountains and embraced the area that lies on both sides of Sulphur Creek, the Geyser Springs, Geyser Peak and most of the country drained by Big and Little Sulphur Creeks. Located in the extreme northeast corner of Sonoma County, the district later became known as the Cinnabar Mining District.[4]

Godwin personally located a number of claims, and men named Southard, Doyle, Kelty, Baxter, Campbell, Palmer, Hayden, Roberson and Van Doren soon joined him.[5] By 1861, some thirty-three thousand feet of claims had been located within the district. Among the claims being worked in that year were the Cincinnati, Dead Broke, Pittsburg, Pioneer and Denver. None of the mines survived the first year, and only the Pioneer was known to have produced any quicksilver.[6] Godwin soon recognized the failure of the small independent mines.[7] In a last-ditch attempt to compete with New Almaden, Godwin consolidated all the claims into one corporation, with himself as president.

Godwin's corporation failed to muster the needed capital and produced no quicksilver during its brief existence. The early mining operations failed primarily because the miners were ignorant of the rather complicated process of reducing cinnabar ore. They only vaguely understood that the ore had to be thoroughly roasted in retorts or furnaces so that the quicksilver

The Mayacamas Range looking south from Pine Flat with Mount St. Helena in the distance. *Courtesy Patrick Dirden Photography.*

vapor would pass off at eighty degrees. The vapor then had to be condensed in large brick condensers, where it came in contact with cold air. Even though they consolidated to procure capital, Godwin and the early miners made no attempts to obtain the equipment for reduction. They apparently were misled by a few deposits of pure quicksilver and hoped that they could obtain the metal by inexpensive placer mining. By 1861, Godwin had given up completely the idea of mining quicksilver profitably. As noted, he enlisted in the Confederate army and was killed in the Civil War. With all the claims abandoned and the leading exponent dead, quicksilver in Sonoma County was forgotten temporarily.[8]

Another significant reason why the Sonoma County mines were not developed more extensively in the 1860s was that the low price of quicksilver made it unprofitable. At fifty cents a pound, even rich deposits did not warrant development. The gold mining boom had subsided, and the New Almaden Mine was quietly producing the greatest portion of quicksilver in the United States. From 1850 until 1874, New Almaden produced 573,150 flasks of quicksilver. This amount enabled the American supply to exceed the demand. In its early days, New Almaden ore yielded a rich 36.74 percent, and it was this factor alone that allowed the Santa Clara

County mine to operate profitably despite the low prices. Quicksilver was not a commodity that attracted those seeking substantial financial reward. Few capitalists relished the thought of investing in an endeavor that had a reasonable chance of being a losing enterprise.[9]

Between 1863 and 1872, the price of quicksilver fluctuated between forty-five and sixty-five cents a pound, but suddenly in 1872, the price began a rapid ascent. By early 1873, it had exceeded one dollar a pound.[10] There were differing and conflicting explanations as to why the price suddenly rose. The gold and silver interests, which needed quicksilver in their mines, screamed that they were the victims of a monopoly that controlled all production and conspired to keep the price high for its own profit. The *Mining and Scientific Press*, siding with the gold and silver people, reported in early 1873 that "the leading cinnabar mines of the world are either owned by or under the control of a gigantic combination of capitalists, and a continued increase in the price of quicksilver may be expected. This combination embraces the control of the New Almaden and New Idria of California."[11]

The same publication named the accused as the "Rothschild-Barron monopolies" and stated that they were located throughout the world and controlled the entire market. There is a great deal of evidence that this "Magic Quicksilver Ring" did, in fact, manipulate the market from 1868 to 1873.[12] According to the *Mining and Scientific Press*, they allegedly were organized so well and maintained such stringent control of the market that it would have been advisable for the Comstock silver miners to break the monopoly themselves. In other words, the silver people were being encouraged to own and operate their own quicksilver mines and circumvent the open market.[13]

There were no known attempts by the Comstock mining interests to purchase and operate their own quicksilver mines. However, if the *Mining and Scientific Press* was a reliable indicator, there was much evidence to suggest that the gold and silver interests fervently hoped that new deposits would lower prices. With the price rising rapidly, the known deposits in and surrounding the Cinnabar Mining District began to attract considerable attention. Numerous claims were relocated, and there was much talk of "mines" in the area. The *Press* hopefully, yet cynically, predicted that if the region "proves as prolific in quicksilver as it is in 'quicksilver mines' we will surely see the speedy downfall of the monopolies."[14]

Despite all the talk about an unscrupulous worldwide monopoly, it appears to have been an oversimplification of a complex set of circumstances. It seems hard to argue with the fact that a small number of capitalists were in the quicksilver business, but with the price of the metal at or near fifty cents, it was imperative that its production be handled by a large, well-financed and efficient enterprise. In addition, quicksilver was only mined in a few places in the world, yet the supply still exceeded the demand in both America and Europe. Only by allowing China to purchase the surplus could the price be kept high enough to make reduction profitable. There is much evidence to suggest that without the high protective tariff on quicksilver, the American quicksilver industry would have perished because the cost of reduction and labor was more in the United States than elsewhere.[15] Relatively few people were engaged in quicksilver production, and their interests were protected by a high protective tariff. With these facts in mind, it becomes obvious that a natural monopoly existed. However, there is no evidence to suggest that this natural monopoly conspired to triple the price of quicksilver within one year.

In 1874, the *Mining and Scientific Press* printed a chart detailing production at New Almaden. The chart clearly showed that the nation's largest producer of quicksilver had seen both the quality and quantity of its ore drop drastically. Its ore had yielded 36.47 percent in 1850, but by 1873 it yielded only 4.87 percent. In 1867, the mine produced 47,194 flasks of quicksilver and in 1873 only 11,042 flasks. In addition, with the Comstock experiencing its last big boom, "the wants of the gold and silver mines are increasing from day to day." The accompanying chart illustrates that California production hit an all-time low at about the same time the price sharply rose. These factors even forced the *Mining and Scientific Press* to belatedly admit, "The price is high from natural causes, and not from the workings of any monopoly."[16]

The realization that no monopoly existed forced the *Press* to embark on another crusade—that of getting the new locations developed. Its editors were well aware that cinnabar was present in the Cinnabar Mining District, but they were not very enthused when quicksilver became the most talked about subject in rural Sonoma County. The gold and silver interests hoped that wealthy investors would flood into the area and immediately set up operations that would rival New Almaden. This, of course, would lower prices. However, there were no willing capitalists, but there was no shortage of adventurous fortune hunters. As noted by the *Press*, there were numerous "honest farmers who dropped the plow and took to the pick."[17] These "farmers" set out to relocate the claims of the 1860s.

Most of the mining activity that had taken place in the Cinnabar Mining District during the 1860s had been on the ledge that runs along the divide

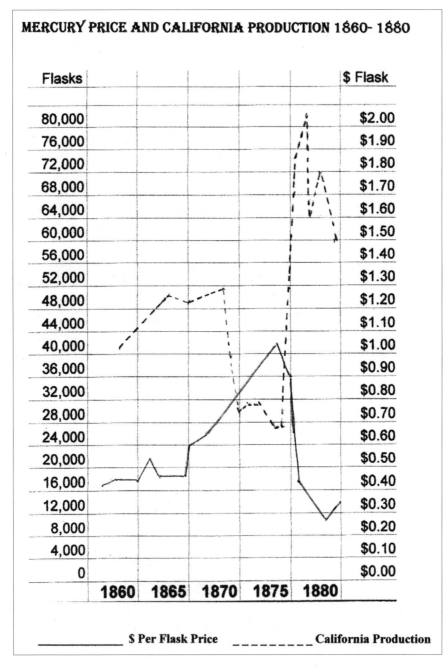

MERCURY PRICE AND CALIFORNIA PRODUCTION 1860-1880

Compiled from California State Mining Bureau *Bulletin 78*.

between the waters of Big and Little Sulphur Creeks. The divide is known as the Hog's Back and was "covered from its base to the crest with scrub oak, manzanita and chemise." The Hog's Back extends from Geyser Peak in an easterly direction toward Pine Mountain and eventually is linked to Mount St. Helena. Local newspapers stated that most travelers into the Hog's Back region were moved by the spectacular scenery that the two-thousand-foot altitude afforded. A witness in 1873 noted that "a beautiful panorama of ever-changing scenery presented itself to view" and that "one can see afar the beautiful Russian River valley with the town of Healdsburg nestled in the shade of numerous groves, surrounded by fields of waving grain and orchards heavily burdened with ripening fruit."[18] On a clear day, one could see far beyond Healdsburg, where the broad Pacific added a blue background to the panoramic view.

While locals were undoubtedly influenced by the positive descriptions reported in the press, it should be noted that, for whatever reasons, journalists routinely chose to print glowing and flowery rhetoric regarding the Geysers and the mining area. There were sparse attempts by newsmen to seek journalistic balance, while firsthand accounts, lacking editorial review, sometimes presented a differing viewpoint. Visitors traversed narrow wagon

The Geysers Hotel in the 1870s. *Courtesy Library of Congress.*

roads that were often hot, dusty and bumpy or cold, wet and muddy, not to mention they were entering the home of numerous rattlesnakes and other undesirable varmints. A differing view was written by James Hutchings, who, instead of "a beautiful panorama" regarding his Hog's Back experience, wrote in his journal, "The ridge much more resembles the back of a horse which has just crossed the plains, or has dieted for some time on shavings, than that of a plump porker."[19]

In the early 1870s, northern Sonoma County was a sparsely populated, agriculturally based region. Most of its inhabitants lived in the Russian River Valley and farmed the fertile lands along the river. With the completion of the railroad as far north as Cloverdale, the valley provided excellent conditions for growing a variety of crops. Both field and orchard crops thrived in the valley, but the surrounding Coast Ranges were of little use or interest to the farmers of the era. Grange activities of the time indicate that the farmers were committed to their agrarian life, since membership in the Patrons of Husbandry soared. The ambitious Grangers established their own bank and co-ops to serve the farmers of the Healdsburg and Geyserville area. The people of the valley had adapted to the trying yet tranquil farm life to which the region lent itself. They were not prepared to cope with or understand the events that were taking place in the mountains to the east of their "fields of waving grain."

The farmers had expressed little interest in the nearby Mayacamas Mountains, but because of early attempts to commercialize the Geysers, the mountainous region was not the remote area it might have been when the quicksilver price skyrocketed. In 1872, the region was accessible by several routes, the best two being the Geyser Toll Road through Healdsburg and the Calistoga Road by way of Knight's Valley and Napa County. Consequently, prospectors and miners had little trouble entering the area in the winter of 1872, despite the seasonal rains. Since traces of cinnabar could be found all along the Hog's Back, there was ample reason for all enterprising fortune seekers to roam through the area in quest of an ore that few of them knew much about. By October 1873, the *Russian River Flag* proclaimed, "The subject of quicksilver mining has become so prominent in this part of the county that the whole region about the geysers has been overrun by prospectors seeking their fortunes in the silvery fluid. Nearly every spot that could be suspected of harboring mercury has been claimed."[20]

A Santa Rosa newspaper reflected this same local fever when it attempted to state calmly, "We have no desire to create a mining excitement," but "during the past summer, in the northeastern corner of Sonoma County, almost at our doors, there has been partially developed the richest and most extensive cinnabar deposit in the world!"[21]

The lust for instant wealth attracted countless prospectors into the district. One must remember that this was barely two decades removed from the California gold hysteria of 1849 and the Comstock silver boom in Nevada with their lingering tales of instant wealth. Accounts of early quicksilver prospecting indicate that prospectors expended much energy while knowing very little about what they were doing. According to the *Napa Register*, "Prospectors are out in force scouring the hills and crawling through chemise brush in search of something rich."[22] The *Mining and Scientific Press* cynically noted that it had witnessed the "landscape dotted with prospectors, each with his little hammer pegging away industriously and enthusiastically at some geological formation imagined to contain cinnabar."[23]

Prospectors overran the area surrounding the Hog's Back and even descended into rugged Devil's Canyon, since a report noted that "its walls now echo the sound of the prospector's hammer, and the prints of his industrious footsteps may be seen in this lonesome retreat, besides that of the heretofore undisputed possessor the grizzly."[24]

One account noted that it was "impossible to estimate the number of people who are 'quicksilver crazy.'"[25] The *Russian River Flag* assured its Healdsburg readers that the number of prospectors was great and that they were not confined to the male population, since "ladies have amused themselves prospecting, and have brought in rich specimens."[26]

Gold and silver folks had wanted avid interest in quicksilver development, but the resulting rush was not what they had in mind. As noted, they had hoped that well-financed and experienced miners would develop the deposits, but "it is not the miners who are looking for quicksilver, but people of all classes have turned prospectors for the time being." The *Mining and Scientific Press* lamented that "red rock or dirt will excite a miner now-a-days as much as a red rag will a bull."[27] Local accounts did little to disprove the notion that most of the prospectors were "greenhorns." A letter to the editor of the *Russian River Flag* reflected this ignorant enthusiasm: "A man here that don't [*sic*] own a quicksilver mine will soon be regarded as an old fogy. Arise, then denizens of Healdsburg who are still mineless, and hie to the mountains! Stake your claims, incorporate companies, issue stock ad infinitum!"[28] With people of this nature making claims, it was no wonder that the gold and silver interests doubted that the new mines would compete successfully with New Almaden.

In contrast to the "greenhorns," there was one organized attempt to prospect the area systematically. Edward Clarke prospected for nearly six months with "an efficient and active corps of quicksilver prospectors" who had "quietly but thoroughly been prospecting the whole cinnabar belt, north of Mt. St. Helena." Clarke believed that by systematic prospecting, he could locate a sizable number of valuable claims. A newspaper account of what Clarke's men had accomplished up to April 12, 1873, included this description: "A contour map of the belt has been prepared, indicating the surface outcrop of the various foundations along its course. This again has been divided into sections, each section being carefully examined, its outcrop, trend, superficial extent and metaliferous character carefully noted and classified for future references."[29]

Clarke was reported to have been satisfied with his work even though it had taken "considerable expenditures of means, and physical labor." Clarke did locate a large number of new claims and hoped to sell them for a handsome profit.[30] The only known sale by Clarke was to the Sonoma Consolidated Quicksilver Company. The price was reported to have been a "valuable consideration." As part of the deal, Clarke later became superintendent of the Sonoma Mine.[31]

In addition to Clarke, there were two other prospectors who seemed to have had an organized plan of what they were doing, even though they admittedly knew very little about quicksilver. The Thompson twins, Granville and Greenville, were two of the earliest locators of claims after the price rise, and they were destined to become the leading citizens of the district. It was while working on the Calistoga Road in 1867 that Granville noticed abundant deposits of red rock that he suspected was cinnabar ore. When the price began to rise, the Thompsons were among the first to locate claims, and theirs proved to be some of the most valuable.[32] Young men of high energy and vision, the Thompson twins' other enterprises in the district will be mentioned later in the text.

By the summer of 1874, seventy-three claims had been located in the Cinnabar Mining District. Most had been located since the price soared in 1872. The *Russian River Flag* compiled and printed a relatively accurate diagram of the new claims that had been located in the district. The diagram clearly illustrated that by 1874, prospectors had left little area unclaimed. While the diagram was detailed, it should be noted that confusion over ownership, some relating to claims in the 1860s, resulted in a number of disputes. Consequently, the *Flag*'s diagram had some discrepancies and omissions. The *Flag* printed the following explanation to accompany its diagram:

> *In the diagram the two rows of words—widening to three and then contracting in an irregular manner to one—give the names of the mining*

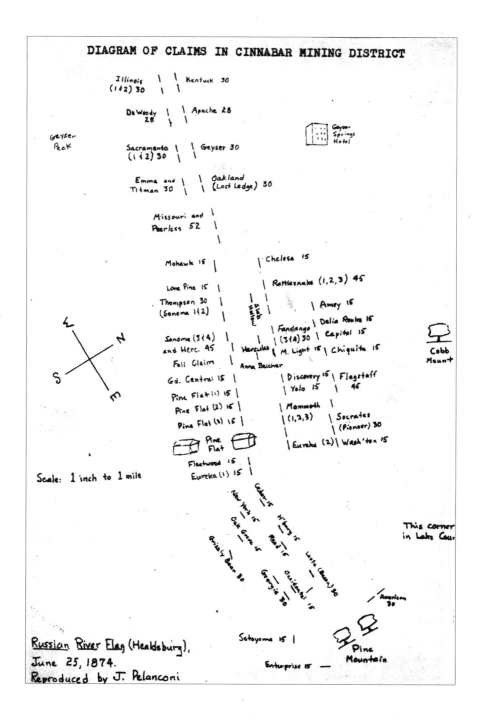

DIAGRAM OF CLAIMS IN CINNABAR MINING DISTRICT

claims. The black dashes show the location and course. When a mine is known by two different names, one of the names is given in parenthesis. The figures in parenthesis indicate the number of the claim. Thus, "Mammoth (1)" shows the claim first located; "Mammoth (2)" is the first extension, etc. The figures outside of parenthesis indicate the number of hundred feet in each claim. Thus "Rattlesnake (1, 2, and 3) 45" shows that three claims, each 1500 feet long (by the customary width of 600 feet) have been consolidated.[33]

The parallel position in which most of the claims were located led many to believe that there was a distinct ledge of ore that ran in a straight line from Geyser Peak in a southeasterly direction toward Pine Mountain. Optimists were plentiful and talked of an unlimited supply of ore running beyond these points. They pointed to the mines that were being developed and worked in Lake and Napa Counties to "prove" that the ledge ran in an easterly direction. The *Russian River Flag* boldly asserted that "it also extends westerly, and has been traced many miles beyond Cloverdale."[34] This proved to be wishful thinking.

A Character of the Time
TIBURCIO PARROTT

Mining booms, like the one in the Cinnabar District, attracted risk takers from all walks of life. There were assorted vagabonds, farmers who dropped their plows, entrepreneurs looking to open a business and Chinese and Mexicans seeking employment—all of whom were hoping to make a quick buck. Not to be left out were well-to-do capitalists, many of whom ventured into the area to survey the quicksilver mines. Most were reluctant to invest in an industry with an uncertain future. A few welcomed the challenge.

John Parrott was a wealthy San Francisco financier and risk taker who, in 1874, invested $52,000 in the Rattlesnake Mine in the Cinnabar District. Previously, during the 1840s, Parrott had been an American diplomat stationed in Mazatlan, Mexico. In Mexico, Parrott accomplished, among other things, the fathering of several illegitimate children. One child with Delores Ochoa, his Mexican mistress, was Tiburcio Parrott, born in 1840. As a youth, Tiburcio lived with his mother in Mexico and was later educated in Massachusetts and England. He eventually followed his father onto the San Francisco banking scene, destined to become an active quicksilver capitalist.

Local newspapers reported that Tiburcio Parrott visited Pine Flat in December 1874. His visit could well have been to check on his father's investment, or perhaps he had an eye on an investment of his own. Whatever his motive, he apparently liked what he saw regarding the potential for financial gain in the quicksilver business. Like his father, Tiburcio proved to be a man unafraid of making decisions. With the aid of his father's

Tiburcio Parrott in his coaching outfit around 1865. Photograph is courtesy of the California Historical Society. Copy was made by Addis & Koch, San Francisco

Tiburcio Parrott in his coach gear. *Courtesy Spring Mountain Vineyard.*

money, he was soon to become the controlling partner of the Sulphur Bank Mine in nearby Lake County.

Parrott's well-managed Sulphur Bank Mine was able to temporarily withstand the downturn in the price of quicksilver and continue operating into the 1880s. The Parrotts had a long history of employing Chinese in their various enterprises, and the Sulphur Bank was known for using significant numbers of Chinese miners. There was evidence that Chinese miners did most of the underground work and, like other quicksilver miners, suffered severe hardships from the effects of mercury poisoning and their work in the hot tunnels of the Sulphur Bank Mine.

An unpublished biography of Tiburcio Parrott detailed how the Parrotts had long employed Chinese workers and were strong supporters of Chinese miners' interests. An 1880 California law prohibited the use of Chinese labor by corporations. The exclusion law emerged from the ugly anti-Chinese sentiment fueled by those who saw Chinese as unfairly competing with American white labor. An independent thinker with strong convictions, Parrott's actions were rarely influenced by conventional thought. In a blatant act of civil disobedience, he insisted that the Sulphur Bank not abide by the law, and his mine continued to employ Chinese. Supporters saw Tiburcio as a champion of Chinese rights, having "taken a cue from Abraham Lincoln." Detractors viewed him as valuing cheap Chinese labor for his personal gain. Leaders in the labor movement called for his hanging. Regardless of his motives, Tiburcio Parrott had taken a courageous stand and was imprisoned for breaking the law and brought to trial. In a landmark case, the Ninth Court of the United States ruled in Parrott's favor, and he was released from prison in March 1880. Succumbing to a sagging market, the Sulphur Bank Mine filed for bankruptcy in 1883.

For the next decade, Tiburcio lived in St. Helena, where he became a legend in his own time. He was closely associated with the Beringer brothers and founded an exclusive wine estate now known as Spring Mountain Vineyard. His numerous investments, many in mining, led to a number of litigations. He was a man of many interests, an art lover and a contributor to various favored causes. A bachelor much of his life, Tiburcio became well known for his flamboyant lifestyle. It is said he regularly patronized the Stone Bridge Bordello in St. Helena, going into town with his servant driving his surrey. Upon entering town, he had the servant blow an English hunting horn to alert the girls that Tiburcio was on his way. In 1880, he built Villa Miravelle, a mansion later used in the 1980s filming of the *Falcon Crest* television show. The beautiful Miravelle landmark is now part of Spring Mountain Vineyard.

No ordinary man, the colorful Tiburcio Parrott strongly influenced both the quicksilver and wine industries, not to mention his social impact on the Napa Valley in the late 1800s. In 1894, he died in St. Helena of stomach cancer. His obituary in the *San Francisco Call* was headlined with "A prince of a Good Fellow is no more." It went on to say that Tiburcio was "generous even to a fault."[35]

Chapter 3

MINES AND THEIR PROBLEMS

By the fall of 1873, eight mines were operating in the Cinnabar Mining District. On the northwest slope of the Hog's Back there were the Thompson Mine (soon to be renamed the Sonoma), Thompson Extension and the Missouri Mine. Farther north and nearer the Geysers, four mines were located in Devil's Canyon, and on the northeast slope of the Hog's Back was the Rattlesnake Mine.

The Thompson brothers had located the Thompson Mine in 1872 but apparently lacked the necessary capital to begin mining. To procure capital, they sold the Thompson Mine to a group of San Franciscans headed by General George S. Dodge for $30,000.[36] They then operated the Thompson Extension, which they mistakenly thought would prove as valuable as their initial claim. The new owners of the Thompson Mine soon incorporated with $3 million capital stock and operated under the Sonoma Quicksilver Mining Company. Besides Dodge, the owners were A.J. Ralston, W.M. Lent and Charles Felton, all of San Francisco.[37] These owners were reported to have been "shrewd, energetic and practical men."[38] They employed fifty men the first summer and were constructing a furnace with a thirty-ton capacity. By November, they reported they had a large amount of ore at their dump and the furnace was to be completed by the last of December.[39]

The Missouri Mine was owned by Litton, Hopkins, Stewart and Elder, all of San Francisco; Morrison of Boston; and McDonald, Graves, Whitton and Keyes, all from neighboring counties. It was located near the eastern extremity of the Hog's Back and near the junction of the Healdsburg and

Calistoga Roads.[40] In 1873, the Missouri seemed to have sold adequate amounts of stock to ensure rapid development. Its owners already had erected a one-ton Roach furnace for reducing the cinnabar ore. A visitor to the mines noted that "there were more evidences of a great and lasting mine in the Missouri than any we noticed."[41] Sufficient capital, from as far as Boston, had gotten the Missouri off to a stable beginning.

The Lost Ledge Mine commenced work in May 1873. Its operators were the original locators, G.L. Cook, R. Stokes, J.H. McCord, George Reeve, J.H. Potts and Calvin Griffith. Located three miles west of the Geysers, the Lost Ledge operated in the summer of 1873 without any capital. However, the owners erected a small retort and were able to pay all their expenses. Because of their lack of funds, an observer noted that "they have been compelled to work economically."[42] There were numerous rumors in the mining region that the owners of the Lost Ledge were negotiating to sell the mine to parties with sufficient capital to work it year round.[43] A newspaper account in May 1873 stated that the Lost Ledge was for sale but for "not less than thirty thousand dollars." With typical journalistic bravado, it then added that this was "a small price for such a mine."[44]

The Geyser Mine did not lack capital. The principal owners were "capitalists of Sacramento City, among whom are Edgar Mills, Geo. W. Gilbert, and E. Motl." Other less important interests were owned by men

A Pine Flat miner, circa 1873. *Courtesy Sonoma County Museum.*

named Baker, Hamilton, Hossack, Ayres, Hardies, Stowell and Dixon, from San Francisco and vicinity. An observer reported that they had very high-grade ore and their financial resources would enable them to "put on a large force, erect reduction works, etc." by the summer of 1874. The same visitor thought "there is every indication of a first rate mine in this place."[45] The Geyser Extension was situated near the Geyser and owned by Clark and Dixon. No one predicted enormous success for the Geyser Extension.

Initial attempts to operate the Kentuck Mine were made by its original locators, Hopkins, Cyrus, Nash, McGarlain, Downing, Hargrave and Boyce. They were all of Napa County and lacked ample capital to operate the mine efficiently. Early indications had seemed to point to a bright future for the Kentuck, but the summer of 1873 had proven to the owners that they could not operate the mine without capital. By November, they were planning to sell at least an interest in the mine to procure financial support. The Kentuck soon became bonded to the more stable Missouri.

It was the Rattlesnake Mine that seemed to cause the most excitement in the summer of 1873. The editor of the *Sonoma Democrat* gave his personal opinion on the Rattlesnake's wealth:

> We have no desire to exaggerate its value, and we do not know how to speak to it without the appearance of exaggeration….Near the surface they struck a mass of cinnabar and vein matter, holding free quicksilver; both vein matter and ore of astonishing richness. The free metal follows the blow of the pick and trickles down the sides in globules.[46]

The owners were John McKay, T.B. Sleeper, H.B. Snow and William Forbes, all of Lake County.[47] They, too, lacked capital and had originally hoped to sell their claim for $150,000, but by the fall of 1873, they had apparently changed their minds, since they had "in sight sufficient wealth to yield them a fortune."[48] Perhaps because of the prospects of a "fortune," the owners soon enticed John Parrott, the prominent San Francisco financier, to buy into the mine for a reported $52,000 and supply needed capital.[49]

The first summer in the mines proved to most observers that there was sufficient cinnabar in the area to warrant further development. The price of quicksilver remained high through the winter of 1873, and the miners talked of the coming dry season when they would construct roads, erect furnaces and procure needed labor. With the price of quicksilver approaching $1.50 a pound, financial success should have been imminent. Yet despite glowing press reports, there appeared to be an air of uncertainly in the mining

An unidentified mine in Cinnabar Mining District. *Courtesy Sonoma County Museum.*

region. Only eight of the seventy-three claims had begun mining, and many of the early claim locators had become discouraged and returned to their agrarian lifestyle in the valley. They had discovered that a handsome financial outlay was needed to begin operations, and there seemed to be too few capitalists willing to supply the needed finances. As a consequence, countless transactions involving claims took place as claim holders tried to find ways to make their endeavor pay. As early as May 1873, a report stated, "Quite a number of transactions are taking place on mining property, transfers of stock, deeds, acknowledgements, etc., sufficient to keep a notary busy most of the time."[50] The busy notary was none other than Greenville Thompson.[51] An April 1874 article in the *Mining and Scientific Press* reflected the continuation of the unstable and nervous ownership:

> *The Lost Ledge Mine has been leased to John A Robertson and Co., of San Jose, for one or two years, with the privilege of buying the mine during the lease at $40,000. The Kentuck has been leased to W.A. Stewart, for $300 per month, during one year, with privilege of purchase during the time, for $30,000. The Rattlesnake having been sold and incorporated, a considerable force has been set to work. We learn from the best authority*

that the sale was without reservation for $52,000 cash. McKay, Snow &
Sleeper, former owners, are still at Pine Flat.[52]

To hinder development further was the fact that the mines were situated in a region that was not conducive to wet-weather operations. Only the Sonoma Mine operated with any degree of success during the winter of 1873–74, and the rainy season put it considerably behind schedule. It had wanted to have its furnace working by the first of the year, but it was not completed until February. The *Russian River Flag* reported on February 12, 1874, that "the Sonoma mine recently shipped forty flasks of quicksilver valued at $4,000, and have an equal amount to ship when the roads will admit."[53] At this time, it was working fifty men and continuing the work in its tunnel both day and night. Most of the other mines in the district were waiting for good weather.[54]

Although it halted mining operations, the wet weather did not dampen local enthusiasm. In April 1874, the *Flag* reported that "the excitement still continues here and many men of means are arriving daily with a view to investing their coin."[55] In May, the *Napa Register* stated, "English,

Work in tunnels was dark and dirty and done with a pick and shovel. *Courtesy Shutterstock.*

New York and Chicago capital, as well as San Franciscan, is now being used freely in the district."[56] Since insufficient funds had been the biggest problem faced by miners in 1873, even these exaggerated accounts were good news. However, wealthy investors already had proven to be hesitant, and they became more reluctant when the following news reached the mining area in June 1874: "According to telegraphic dispatches, the bill extending, until the 1st of January next, the time for expenditure of labor on mining claims located previous to May 10th, 1872, has been signed by the president and become a law."[57]

Since most of the Cinnabar District claims were originally located in the 1860s, this extension gave the old claimants who had given up their claims an additional chance to reestablish their right to the claims. The *Napa Register* gave its straightforward opinion on the extension:

> *The mischievous effects of this rascally legislation are already apparent in our immediate vicinity. Several of the best quicksilver mines—locations which adventurers had squatted on 13 years ago, and totally abandoned for years are now suddenly claimed under color of it provision, after others have gone in and by the expenditure of their money and labor and made them valuable.*[58]

The *Mining and Scientific Press* gave this account of the confusion that ensued when it stated that "the Hog's Back mines have been disturbed of late by men from different parts of the country coming on and jumping the Missouri, Lost Ledge, Geyser and Kentuck."[59] Miners have traditionally reacted fervently to the slightest hint of claim jumping, and Sonoma County miners proved not to be exceptions: "An attempt was made to jump the Sonoma by a party of old claimants; while another tried to do the same with the Flagstaff mine, but with the help of about a dozen Henry rifles and several six shooters they were at last persuaded to relinquish their efforts."[60]

Most of the incidents, however, eventually were settled in spectacular court battles that had the profound effect of exciting the Sonoma County populace but did little to enhance the interest of potential investors. The Emma and Titman, Geyser, Missouri, Kentuck, Lost Ledge and Pioneer were soon tied up in lengthy and costly litigation. In June 1874, the Missouri, Geyser and Lost Ledge decided to pool their resources and fight the old claimants jointly with top-flight lawyers. Their lawyers immediately assessed the mines "from $10,000 to $20,000 (as emergencies demand)" in order "to defray expenses in defending their suit."[61]

Consequently, local journalists reflected wishful thinking rather than factual knowledge when they stated that capital was readily flowing into the district. "Men of means" may have been arriving daily, but they were not "investing their coin." The *Mining and Scientific Press* lamented the lack of capital in the district and felt that the "apathetic capitalists" should have been willing to take the risk that litigation afforded. The paper noted that Sonoma County miners "have spent months in this city [San Francisco] endeavoring to get capital to aid in opening these mines, and most of them have gone back disgusted." In conclusion, the *Press* noted, "there isn't any capitalist in this city that wants to go into quicksilver mining, unless they could buy a New Almaden."[62] In addition to their court problems, the Sonoma County mines had convinced conclusively only the local "greenhorns" that a true bonanza existed in the Hog's Back region. Most men of means were seeking safer investments.

In spite of the fact that ownerships were in flux and additional funding was not readily forthcoming, the established mines reopened, and numerous others were hoping to commence initial operations by April 1874. The Missouri Mine was the first to resume operations. John Harris, formerly of New Almaden, was now superintending the work on the Missouri.[63] The Emma and Titman Mine, owned by "Messrs. Ball & Whitehead, Dr. Stillwagon and Hon. W.W. Pendegast," looked to be one of the most promising new mines since its owners reportedly had money to invest. They had hired William W. Whitton as superintendent. Other mines that by the fall of 1874 had hired men and begun the relatively lengthy process of setting up operations included the Anna Belcher, Georgia, Occidental, Healdsburg, Bacon, Flagstaff, Chiquita, American and Fleetwood. None of the new mines had any recorded production in their initial year, but there is evidence that they sold some ore to the established mines for reduction.

In the fall of 1874, the Oakland (Lost Ledge), Rattlesnake, Sonoma and Missouri were shipping metal every week from furnaces in full operation. A newspaper account predicted that the Geyser, Flagstaff, Anna Belcher and Georgia would soon commence similar shipments "if the work in progress is not retarded by inclement weather." The price of quicksilver had dropped from its high of $1.65 a pound, many of the mines were facing prospective litigation and the rainy season was approaching. Nonetheless, the *Sonoma Democrat* predicted, with characteristic optimism, that "the mines of the county will furnish, by spring, from seven to eight hundred flasks of metal a month."[64]

Despite blind optimism, the fact remained that in a year, only four of the seventy-three claims had been able to produce any quicksilver, and two

Mines were located on hillsides in rough terrain. *Courtesy Shutterstock.*

of the producers were in financial difficulty. The Oakland, Rattlesnake, Missouri and Sonoma Mines were the only ones that had been able to procure the bare necessity of a retort or furnace for reducing the ore, and they had accomplished the feat only with a maximum amount of difficulty. Besides the significant cost, the miners discovered the problem of getting the cumbersome material for furnace construction from the valley to the mines. Most components for reduction works came from San Francisco and were shipped northward by rail. Both Calistoga and Healdsburg were served by rail, but both communities were a considerable distance from the mines and linked to them by narrow, winding and relatively steep wagon roads. One of the numerous near-disasters in transporting reduction equipment to the mines was noted by the *Napa Register*: "Foss and Connolly left on Tuesday last with three large boilers to be used as retorts for the Sonoma mine. Everything passed off smoothly until they got up above D. Fairfield's on the Geyser road, when one axle gave way and one of the huge boilers was allowed to rest on terra ferma."[65]

One significant problem that the miners faced successfully was the matter of getting brick for their furnaces. Since a furnace with a twenty-

Mining and Scientific Press, August 1874.

ton capacity would need at least 100,000 bricks, it was not feasible to transport them from the valley. The Rattlesnake hired a Mr. Leonard from San Jose to burn 100,000 bricks near the mine, while the Thompson brothers contracted with Dick Swift and Thomas Simmons of Calistoga to furnish 60,000 for the fifteen-ton furnace at the original Thompson Mine. Simmons and Swift set up their kiln at Pine Flat in the midst of the mining area. The Reverend J. Daubenspeck later started a similar brick-making enterprise in Pine Flat. The brick-makers obtained the necessary commodities for the brick, such as clay, from the nearby countryside and produced a brick of relatively good quality.[66]

The Geyser, Anna Belcher and later the Cloverdale Mines installed Knox and Osborn furnaces. The Rattlesnake and Sonoma Mines used Lockhardt furnaces, which were similar to the Knox and Osborn. The Missouri and Oakland Mines reduced their ore in smaller retorts, which proved to be less practical than the furnaces. The furnaces were run day and night and burned up to a cord of wood in twenty-four hours. The furnaces were reputed to have been able to burn anything from small chaparral to logs twenty inches in diameter. The firewood generally was cut into four-foot lengths and hauled to the mines in narrow wagons with high sideboards.

In January 1874, the *Flag* stated that 300 to 400 miners were employed in the district.[67] By fall, the number had risen to an estimated 1,000 miners "working or prospecting in and about Pine Flat."[68] At the time, there were six established mines employing at least several dozen men each and about fifteen more mines employing a small crew trying to commence operations. In June 1874, the Rattlesnake employed 70 men while the Lost Ledge, Missouri and Sonoma used about 40 each.[69] By September, all had increased their labor forces, with the Rattlesnake employing 120 men.[70] The men were employed at a variety of tasks, with relatively few of the "miners" actually working in the mine tunnels. Although the construction of tunnels and the extraction of ore were of prime importance, others needed to be engaged in building roads, procuring tunnel timbers and securing firewood for the furnaces.

Obtaining an ample supply of both firewood and timber was a necessary and difficult prerequisite to successful mining. Firewood had to be obtained in large quantities, since a furnace often would burn as much as a cord of wood in twenty-four hours. The topography of the area did not lend itself to the easy procurement of wood; the mines were generally situated on the chemise-covered ridges, while firewood and timber had to be sought in the rugged and brushy canyons. The Rattlesnake had to build a road of over two miles to a timber stand.[71] The building of these roads into the steep canyons was a painstaking endeavor. The superintendent of the Missouri noted, "We have, under most trying circumstances, built a road over a mile through the thickest of brush and boulders, up the little Sulphur creek to our wood

Ore-bearing rock was transported to furnaces by wheelbarrow and ore cart. *Courtesy Shutterstock.*

land."[72] Without firewood, the furnaces could not be operated, but firewood was relatively easily found, oak and madrone being common in the canyons. More difficult to obtain were the pine and redwood timbers used in the tunnels. The Missouri superintendent gleefully reported that he had "a force of men cutting timber, and now have ample supply, and our tunnels are undergoing repairs, of which heretofore they of necessity have been left in a dilapidated condition."[73] The Missouri was not atypical, since most of the mines employed a large percentage of their labor force outside the mines.

Labor was not particularly scarce in Sonoma County during the 1870s, but there was a definite shortage in the mines. A report in early 1874 stated that "laborers are in demand and there is no need for idleness in this district."[74] Because of the harmful effects that result from direct contact with quicksilver (often called salivation), few white laborers were willing to get anywhere near the metal. Consequently, most mines employed Chinese and Mexican help even though some stated they would have preferred all-white labor. Many of the Mexicans had experience with quicksilver at New Almaden and did not seem to fear salivation. The Chinese simply were not told of the danger, and therefore, most were not leery of the possible consequences. The Chinese would also work for less pay but were considered to be less intelligent and untrustworthy. There is no evidence to substantiate these claims, and it seems feasible that the Chinese were the victims of the prevailing anti-Chinese sentiment that was sweeping California in the 1870s.[75]

The Rattlesnake Mine was the largest employer of Chinese in the Cinnabar District. The Rattlesnake superintendent, B.F. DeNoon of Virginia City, personally took fifty-three Chinese to the mines in April 1874.[76] Unlike the mines of Napa County, the Rattlesnake was the only significant Cinnabar District mine that employed considerably more Chinese than white labor. Most of the mines had nearly equal numbers of white and Chinese workers. Only the tiny Fleetwood mine boasted that it had "nothing but white labor." The American had nearly all "Mexican miners and a few Chinese and white."[77]

Work in the mines was dangerous, both from the threat of accidents and from the effects of mercury poisoning. At the time, there was precious little mention of the toxic effects of contact with mercury. Most suffering from mercury "salivation" were Chinese and Mexican miners and warranted little ink in the local press. While the symptoms of mercury poisoning were obvious and well known, as evidenced by the Mad Hatter of *Alice's Adventures in Wonderland* notoriety, there were few precautions taken at the mines.

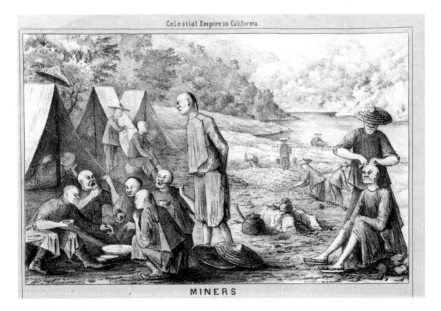

Celestial Empire in California.

MINERS

Chinese miners lived in ramshackle shacks and tents near the mines and outskirts of Pine Flat. *Courtesy Library of Congress.*

Mine tunnels were constructed into hillsides, some as deep as two and three hundred feet. Generally, the tunnels were seven feet high and about five feet wide. Ideally, the tunnels were level with a slight ascent to facilitate the draining of any water and also to make it easier to remove ore and earth. The tunnels required timbering for support, and the local oak, fir and pine trees were used for that purpose. Shafts were often dug down into the earth for the purposes of prospecting. None of this was easy or safe work. Miners in nearby Lake County reported that they experienced horrid temperatures in the tunnels from the nearby thermal activity. Conditions were most likely similar in some Sonoma County mines.

One report in June 1874 noted that "Martin Oates, a miner in the Missouri mine, near Geyser Peak, while timbering in a dangerous place in the upper tunnel, was killed by a large bowlder [sic] that fell upon him." Oates was eulogized as having had "quiet and industrious habits, leaving a widow in Cornwall, England."[78] Many of the tunnels encountered solid rock, and dynamite blasting was common and dangerous. In July 1874, "a careless blast from a new mine within one-quarter of a mile of Pine Flat, sent a large rock crashing through the side of the Thompson Hotel one day last week, barely missing several bystanders."[79] One can only speculate as to the number of unreported serious maladies, injuries and perhaps deaths suffered by Chinese, Mexican and other miners.

Wages for white miners were usually $3.00 a day, while the Chinese were paid from $1.25 to $1.50 a day.[80] Several of the mines provided accommodations for the white miners. Board was usually $5.00 or $6.00 a week, while the Chinese provided their own board. The Missouri was the first in the district to build a boardinghouse. In June 1874, the *Flag* reported, "Ben S. Bradshaw keeps a boarding house at the mine and has twenty-five boarders at five dollars a week."[81] The Oakland Mining Company (operated by the Lost Ledge and Flagstaff claims) built a two-story boardinghouse at the Flagstaff.[82] After the community of Pine Flat was constructed, many of the miners began to live in the town and commute the short distance to the mines.

It had taken more than a year of preparation, but by 1874, the mines were employing a sizable labor force and extracting ore in measurable quantities. Sonoma County's quicksilver production in 1874 (most from the Cinnabar District) totaled 1,700 flasks, in contrast to the 50 produced in 1873. With one full flask weighing seventy-six pounds, the total value of the 1,700 flasks was $178,806, which is a small figure when one considers the number of mines that were hoping to show a profit. In 1875, county

Timbers to brace tunnels were cut in nearby canyons and hauled to the mines. *Courtesy Shutterstock.*

production dropped off to 1,218 flasks.[83] The production figures indicate that none of the mines had proven to be bonanzas, but the local optimists hailed the Sonoma Mine as the "banner producer of the district" with "$200 per day of metal."[84]

Even though the mines were finally producing significant quantities, experienced miners found little reason for jubilation. The mines had been in operation for over a year in an area that undeniably contained cinnabar ore, yet they had relatively little to show for their efforts. Sonoma County's 1,700 flasks in 1874 did not rival New Almaden, even though New Almaden's 1874 production of 9,084 flasks was its lowest output since 1860. The *Mining and Scientific Press* attributed the sluggishness of operations to two factors. The *Press* acknowledged the fact that many of the "developed mines are lying dormant under the ban of litigation, or are worked slowly under injunction." In addition to hindering progress, the legal entanglements discouraged the influx of needed capital. However, the *Press* cited another reason that it considered to be equally significant: poor management.

Quicksilver was weighed and poured into individual casks weighing seventy-six pounds each. *Courtesy Santa Clara County Parks and Recreation Department.*

According to one report, the mines "are owned or managed by greenhorns. Deep and expensive shafts are sunk off the lead, and on the wrong side of the dip. Tunnels are being run from a long way off the lead to strike it at a good depth on its broadside, when tunnels could be run end on from the same depth."[85]

The management was allegedly so bad that some of the mines were developing slowly because they were forced to use "the muscle and intelligence of the children of the flowery land [Chinese], intelligent and skilled white labor declining to work under their management."[86]

From the point of view of the mining establishment, the shoddy management of the Cinnabar District's mines was the inexcusable factor in the slow growth and progress. The California State Mining Bureau tactfully but plainly stated that "much injudicious work was expended in these locations."[87] A reporter for the *Mining and Scientific Press* echoed similar sentiments in much more explicit terms and noted, "I have been in mining districts or rock camps, but have never known one to have so many impediments to its development and prosperity as this of Pine Flat."[88]

B.F. DeNoon, who was a veteran of Virginia City mining, ran the Rattlesnake Mine. The *Russian River Flag* chose not to directly denigrate the

An abandoned mine in Devil's Canyon near the Geysers. *Courtesy Healdsburg Museum.*

greenhorns but instead featured DeNoon in an article as a superior mine operator. Of the Rattlesnake Mine, the *Flag* wrote:

> *Its keeper is a warm-blooded, keen-eyed, tempestuous old grizzly, full of vim and enterprise, and impatient at the sight of our Sonoma county sluggishness…if two dozen such men would come into these Sleepy Hollow precincts of our county and infuse their spirit into such of our people as need it, and show them how to do business on business-like principles, we should soon be on the up-grade to a condition of prosperity far higher and greater than Sonoma county has ever known.*[89]

Local newspapers were universally (some would say blindly) supportive of local miners, yet the quote by the *Flag* was certainly a thinly veiled and severe criticism of overall mine management in the Cinnabar District.

While the mines were developing slowly under greenhorn management and most did not have the experienced leadership of a B.F. DeNoon, one must question the harshness of the accusations of mining industry voices. As noted, there were certainly other factors that slowed progress and made delays imminent. Additionally, quicksilver mining and processing was

The *Russian River Flag* office in Healdsburg. This was often called the Grangers building. *Courtesy Healdsburg Museum.*

not a widespread endeavor and certainly not as simple as placer mining for gold. There was not an abundance of quicksilver mining experts with management skills. It is also unlikely that experienced and skilled laborers were as readily available as the *Mining and Scientific Press* implied, especially in the relatively inaccessible reaches of northeastern Sonoma County. The reference to unskilled Chinese labor probably had more to do with prevailing racial bigotry than actual fact, since Chinese laborers had successfully mined gold, built railroads, mined quicksilver in other areas and proven to be a tireless and successful source of labor.

A Character of the Time

CUM YUK

Imagine—and imagine you must, since no known firsthand accounts, place names or monuments exist—hundreds of Chinese laborers in the Cinnabar Mining District during the summer of 1874. With pigtails down their backs and wearing basket hats and blouses and pantaloons, Chinese men quietly toiled in Sonoma County, living and laboring under primitive and often dangerous conditions. They tirelessly built roads and worked the quicksilver mines, all too often innocent victims of prejudice, violence and mercury poisoning.

Most Chinese laborers were men who came to California seeking gold and later to work on the railroads that were built in the 1860s and early 1870s. Most of Sonoma County's Chinese quicksilver miners came out of San Francisco, arriving in Healdsburg by rail. These new arrivals were then transported by stagecoach from Healdsburg to the mines. Mine owners often shipped veteran Chinese miners, sometimes described as "sick toothless wrecks," out of the area on stagecoaches by way of the Calistoga Road. This was to avoid any contact with the new arrivals from Healdsburg.

The humble Chinese had a culture and language far different from the English-speaking Americans living in Sonoma County in the 1870s. Culturally different and thoroughly misunderstood, the Chinese experienced blatant prejudice and occasional violence. They found it particularly difficult to make the transition from temporary laborers to established residents and community participants. Some Chinese merchants operated stores that served the Chinese populace. However, it was Chinese laundrymen who typified Chinese businessmen of the time and hinted at token acceptance and assimilation. Absent significant capital investment, laundries required only ambitious laborers willing to work long hours. These conditions were well suited to the impoverished and ambitious Chinese immigrants. In addition, laundries were a service not provided by white businesses.

The most noteworthy Chinese businessman in Healdsburg during the 1870s was Cum Yuk. While Sonoma County had its share of blatant anti-Chinese sentiment, the local press and some entrepreneurs, eager to employ cheap Chinese labor, provided a degree of respect for a few local Chinese businessmen, and this included Cum Yuk. Like all Chinese immigrants, Cum Yuk could not vote or testify in court against whites, but he did provide needed services and was accepted as part of the Healdsburg business community. His laundry, Pioneer Chinese Laundry, ran advertisements in the *Russian River Flag*:

CHINESE LAUNDRY.
CUM YUK—*Proprietor.*

CLOTHING WASHED AND IRONED IN *the neatest and best manner, with the least possible amount of wear, and at reasonable rates. All packages of clothing received at the residence of patrons and delivered again without loss of pieces.*
Healdsburg, July 24, 1873.

At some point during the quicksilver rush, Cum Yuk opened a store at the Geyser Mine near Pine Flat, where he undoubtedly sold goods to the Chinese miners. It was reported that he sold the Pine Flat store in 1875 and returned full time to his laundry in Healdsburg.

Employing a number of Chinese immigrants who worked twelve-hour shifts, Cum Yuk ran a thriving business, having purchased his laundry from Hop Sing in 1871. A typical Chinese laundry had ironing boards lining the side walls, with wash tubs in the back of the room. When the clothes were washed and the water pounded out of them, they were hung on the lines outside at the rear of the building. Hand irons were heated on a charcoal stove in the middle of the room. With both hands occupied holding and ironing the clothes, workers would fill their mouths with water and spurt a spray to dampen the clothes while ironing.

Cum Yuk undoubtedly provided regular customers personal service and clean, well-ironed clothing with stiff shirt bosoms, cuffs and collars. As with most Chinese laundry owners, he spoke some English and acted as a go-between for the two divergent cultures in Healdsburg—and for a short time in Pine Flat. Chinese laundry owners were often well liked in the community. An April 1873 article in the *Russian River Flag* entitled "An Intelligent Chinaman" reported that "Cum Yuk, who is 'boss' washerwoman [*sic*]...is progressing quite rapidly in mastering the English language" and "is courteous and gentlemanly, and is rapidly adopting American ideas of living."

While some Chinese were adopting "American ideas of living," they were not immune to the prevailing anti-Chinese sentiment. One need look no further than Dennis Kearney's visit to Healdsburg for an example of the hostile environment Cum Yuk and other Chinese experienced up close. In 1875, the firebrand labor leader from San Francisco brought his "The Chinese Must Go!" rhetoric to a rally in Healdsburg's Plaza, only a stone's throw from the Chinese Laundry. Violence against Healdsburg's Chinese was accepted, and the town later passed an ordinance prohibiting laundries from doing business within the city limits. The 1880 statewide Chinese Exclusion Act suggested that assimilation was not easy for the Chinese. However, in spite of the open hostility, evidence suggests that Cum Yuk was a decent, honest and respected man, able to tolerate the abuse, co-exist and thrive as a Healdsburg businessman.

Cum Yuk's 1874 marriage in San Francisco was duly reported in the *Russian River Flag* as "Cum Yuk, of Healdsburg, to Miss Chun Loi, late of the Celestial Empire." His eventual departure and move to San Francisco was reported with a hint of sadness. Most local journalists did not promote anti-Chinese hysteria, yet hundreds of Chinese came and went through Healdsburg with nary a mention of their personal lives in the newspaper. By competently providing a needed service, Cum Yuk attained a positive status totally unfamiliar to most other Chinese immigrants in the area. An 1876 news article, in what was a bit more than faint praise, reported, "Cum Yuk is the best Chinaman we have ever known." The article spoke of his "excellent character." At the very least, Cum Yuk was an honest man and perhaps shrewd enough to know that by tolerating the prejudice and purchasing ads in the local papers, he would increase business, enhance his standing in the community and endear himself to the press. Whatever his motives, Cum Yuk was a wise and worthy man—a character of the time.[90]

Chapter 4

SONOMA COUNTY'S
REACTION TO ITS MINES

E ven though the mines were staggering on the brink of disaster, only mining experts ascertained the gravity of the situation. There were few, if any, mining experts residing in the valley communities that served as outlets for the mines. Encouraged by a sympathetic and myopic press that sat poised to grasp the whisperings of any new development, mining excitement reached a fever pitch in both Healdsburg and Calistoga. Newspaper accounts indicated that the fever also affected the Sonoma County communities of Geyserville and Santa Rosa, as well as Napa and St. Helena in Napa County. The excitement was best exemplified by a frantic attempt by Sonoma County to build a better road to the mines.

Not long after the mines opened in 1873, a Santa Rosa newspaper noted, "The mines will add largely to the taxable wealth of the county and offend the local market for our farm products."[91] As noted, there were two existing routes to the mines, one from Calistoga and the Geyser Toll Road from Healdsburg. The Calistoga route was shorter and had a better grade than the other route, but the Santa Rosa paper suggested that "both Healdsburg and Cloverdale can, at small expense, be connected by good roads with the mines" because "they are nearer those two points of supply than to Calistoga."[92] Because it was the nearest sizable community, Healdsburg appeared to be the natural outlet. A San Francisco newspaper reported, "Healdsburg is destined to become the outlet for a large number of mines."[93] Healdsburg, however, did not have the better road.

Geyser Road on the bank of Sulphur Creek. *Courtesy Healdsburg Museum.*

Geyser Road from the top of the ridge looking eastward toward the Geysers. *Courtesy Healdsburg Museum.*

With most of the trade heading toward Calistoga, Healdsburg soon became acutely aware of its dilemma. While entertaining frightful visions of Calistoga as a wealthy mining terminus, the *Russian River Flag* embarked upon a crusade to have the needed road built immediately. As early as October 1873, three months before any quicksilver was mined in the area, editorials of the following nature were appearing in the *Flag*:

> *The distances from the mines to Calistoga are from 20 to 25 miles, and to Healdsburg from 14 to 18 miles. It is clearly in the interest of the people of this town to see that not only the road is speedily built, but every other road that may be necessary to secure the trade of these mines. The distance to a trading post is not so much a favor in Healdsburg that we can afford to leave the matter to adjust itself. Calistoga already has the advantage in the fact that most of the travel to and from the mines goes through that town, and the people of that place will make every exertion to keep the travel and trade in their favor. The mines are located in this county; the miners will vote in this county; they must go to our county seat to do their recording, and if our people show liberal and enterprising spirit they will prefer to do all of their business in this county.*[94]

The *Flag*'s editor cannot be accused of making any erroneous statements, but it should be noted that he had determined the worth of the mines long before mining experts cared to speculate. He also had secured the favor of his hometown advertisers.

By November 1873, the citizens of Healdsburg had called a town meeting in the Sotoyome Hotel to "consider the propriety of building a road from Healdsburg to Pine Flat." H.M. Wilson was chosen chairman of the meeting, with I.N. Chapman serving as secretary. The group adopted the proposition that "a committee office be appointed, to view out the best route for a road." The committee was composed of prominent Healdsburg businessmen Wood Bostwick, H.B. Snow, I.N. Chapman and T.W. Hudson, who had also served as a state assemblyman. The fifth member was Granville Thompson, who had his own peculiar interest in a better road to the mines.[95]

The meeting was held on a Saturday night, and the next Tuesday, the committee headed toward the mines in search of the "most direct route possible." The next Saturday, it presented the following report at the town meeting:

> *We find that the direct route from Healdsburg to Pine Flat (which is regarded as headquarters in the mining district), is the best ground to build*

a road on. This route will run in a northeasterly direction from town,
passing near Silas Rodgers' farm, crossing Russian River near Soda Rock.
After crossing Alexander Valley nature has opened a way into the mountain
through the canon [sic] of Sausal Creek. This creek heads near Pine Flat
and runs directly towards this town, making this a natural and easy way
for a good road.[96]

The committee estimated the cost of the road was "not to exceed
$2,000, exclusive of survey."[97] The report was adopted, and the chairman
appointed another committee to solicit subscriptions and employ an
engineer to survey the proposed road and set the grade stakes. The
community had definitely taken the offensive in attempting to obtain its
share of the anticipated mining wealth.

A surveyor named C.B. Thomas was employed immediately and had
a preliminary report in before the end of November. He found that the
Sausal Canyon route was the most practical and that a road through that
canyon would be about sixteen miles in length, which would be five or six
miles shorter than the Calistoga route. Thomas added that the steepest
grade would not exceed sixteen inches to the rod, which would make heavy
freighting feasible. Thomas's estimation of the cost was at $6,000.[98] The
committee was satisfied with the report and not dissuaded by the estimated
cost. It authorized Thomas to make an extensive and detailed survey of the
proposed road.

Thomas immediately returned to the mountains. By mid-December,
he had completed his work and presented his findings to the Healdsburg
populace at still another town meeting. Thomas had started his survey
in Pine Flat near where the route from Calistoga entered the Flat. From
there, he followed the ridge that separated Knight's Valley from Sausal
Canyon and finally dropped off into Sausal Canyon and followed along
the left bank of Sausal Creek. He terminated his survey at the point where
Sausal Creek entered Alexander Valley.[99]

Thomas found that this section of the road through the mountains could
be no shorter than 8³⁄₁₆ miles. Since the difference in elevation from Pine Flat
to the valley was a significant 1,920 feet, any shortening of the road would
make the grade too steep for heavy freighting. Thomas reported that in going
from Pine Flat to Healdsburg, the entire route would be a downgrade, with
the average grade a little less than nine inches per rod. The report stated that
only about 2 miles of the route was covered by brush and timber, of which
only about a half mile was thick. There were four ravines and forty-one

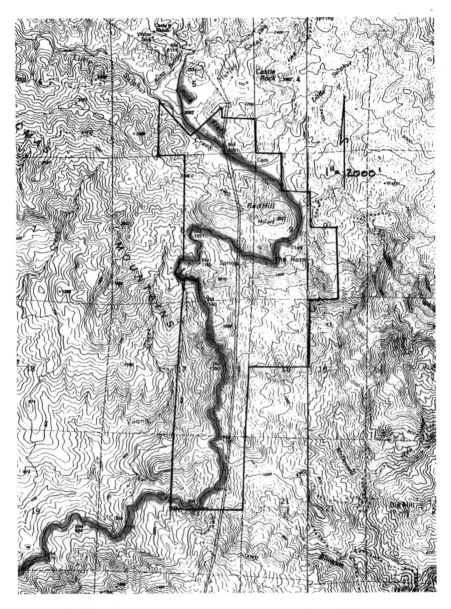

Sausal Road route map, taken from *Historical Atlas of Sonoma County, 1880*.

gulches that had to be crossed. The road itself was to be 12 feet wide with turnouts at least every two hundred yards.[100]

Healdsburg's zeal for a road that was shorter than the Geyser Toll Road and would compete with the Calistoga Road sparked a colorful series of articles between the *Napa Register* and the *Russian River Flag*. Writing under the nom de plume of Bret, the *Register* editor stated he spent several weeks in Pine Flat and studied the road possibilities. Calling the proposed Sausal Road the "forty gulch route," he proclaimed the road would need to be longer than proposed in order to have a grade that would allow for hauling freight. According to Bret, "Any one that has any judgment about such things can tell at a glance that a practical road could never be constructed on that route." The *Flag* referred to Bret as the "Calistoga wiseacre" and quoted Granville Thompson as saying he had never seen Bret in Pine Flat, opining that "Bret's gas about our forty-gulch route…must have been in a spiritual trance." The *Flag* advised Bret "to take lessons in truth-telling, geography and exact measurement."[101]

Healdsburg had taken the initiative in proposing a road to the mines, and few citizens argued with the fact that a better road would benefit the community, but when the question of finances arose, friends often parted company. The citizens soon became divided into factions concerning the road to Pine Flat over both the route and who should foot the bill. The *Russian River Flag* led the faction that clamored for full speed ahead on the Sausal route without much regard to who paid for the construction. This group hoped that the county would maintain it. The main advocates of this route, besides the *Flag*, were Hudson, Wilson, Chapman, William Reynolds, George Jacobs, G.W. Burkus and, of course, the Thompson twins. Opponents of the Sausal route pointed out that it was not surprising that these men favored public financing of the road, since most of them were either miners or Healdsburg businessmen who had the most to gain from such a road.[102]

In leading its faction, the *Flag* was quick to point out that if Sonoma County was reluctant to aid the mining interests, Napa County would certainly not hesitate. "It happens to be Sonoma's good fortune to have rich quicksilver mines discovered within her borders, but her ambitious neighbor, Napa, claims them already as 'our mines.'"[103] The *Flag*'s fears do not seem to have been unwarranted, since articles calling for improvement of the Calistoga route were common in the *Napa Register*: "The people of Sonoma County are straining every nerve to draw the trade in that direction, and will leave no stone unturned to secure it. We have the only natural outlet from these mines, and by the use of a few thousand dollars, we can secure this trade for all time to come."[104]

The *Napa Reporter* echoed similar sentiments in saying that "trade and traffic, if secured to Napa Valley, would give a wonderful stimulus to do business in Calistoga. It is a manifest that the prize is worth contending for."[105] However, despite journalistic prodding, Napa County residents proved reluctant to "strain every nerve" to secure the trade of the unproven mines. Besides, they were getting the bulk of the trade by doing nothing.

Leading another faction of Healdsburg road enthusiasts was the volatile Healdsburg lawyer Captain Lewis Adelbert Norton, a man of self-importance and known as an outspoken and shrewd businessman. In 1859, Norton had been appointed by the governor as captain of the Sotoyome Guard, a local militia group that never procured arms and was disbanded in 1861. Nonetheless, Norton was a respected leader, and he and his group of followers thought it impractical to build a road through Sausal Canyon when the Geyser Toll Road headed in the same general direction. According to them, a route "with as good a grade, could be had by following the Healdsburg and Geyser Toll Road to a point about thirteen miles from Healdsburg and going up the Little Sulphur Creek."[106] This smaller section of new road was said to cost the taxpayers only about $1,600, which would have been a handsome savings.

However, Norton was a primary owner of the Geyser Toll Road and may have been considering more than the taxpayers' financial welfare. The Geyser Toll Road was originally built with private funding to compete with the Calistoga route to the Geysers. According to the *Napa Register*, "They [Norton and partners] had expended a large amount of money" and "found they had got a white elephant on their hands" with a route that "is so steep that it takes two *good* horses to pull two small men over the road." Nonetheless, Norton bitterly opposed the Sausal route and took his proposal to the board of supervisors, suggesting the county buy the Geyser Toll Road and declare it a public highway. Norton claimed the Geyser Road cost $13,000 to build, but he "would sell out to the people at a liberal reduction, say $4,000." The *Register* suggested that "if Captain Norton can persuade the Board to pay him and his associates $8,000 [*sic*] for this elephant, they will be considered smart by some, but not by the taxpayers." Not met with enthusiasm by the board, Norton then asked only that it examine his route. If it proved impractical, he agreed to support the Sausal route.[107] Since Sonoma County newspapers were on record as supporting the Sausal route and perhaps skeptical of the motives of Norton and his followers, they gave it little coverage. Having both proposals before them, the board of supervisors discarded Norton's road in favor of the Sausal route.

As one might expect, the question of who should finance the road led to considerable debate. Many people in Healdsburg felt that even

A group of 1870s Healdsburg businessmen with banking, mining and road building interests. *Left to right*: E.H. Barnes, Ransom Powell, L.A. Norton and R.H. Warfield. *Courtesy Healdsburg Museum.*

though Healdsburg would be the immediate outlet, there would be other benefactors, Santa Rosa being a notable example. This notice appeared in the *Flag*: "Everybody in this part of the county is signing a petition asking the County Supervisors to declare the proposed road between Healdsburg and Pine Flat a county road. The county at large needs this road and the building of it is properly a county charge."[108] Santa Rosa papers joined in the discussion and continually clamored for a road, but it is interesting that they never mentioned a petition suggesting that they aid in financing it. Sonoma County at large fervently desired access to the mines but not badly enough to pay for it.

The mines had hinted earlier that they might help with the road. Many citizens of Healdsburg considered this only reasonable, since they would definitely benefit from its construction.[109] However, when it came time to talk of finances, the mines balked at the idea of financial contribution. Edward Clarke, the superintendent of the Sonoma Mine, prepared a report that was read at a Sonoma County Board of Supervisors' meeting. In it, Clarke stated succinctly the position of the miners:

Gentlemen, we entrust you to give us good roads and we will exchange our silver and gold for your valley products, the wealth of the mountains will flow into and enrich the citizens of the valleys, and will be mutually benefitted. The Russian River country is the natural outlet from the mining district; but should you fail to recognize the importance of this valuable mining district and refuse our reasonable demand for good roads and connect us with your base of supplies, the valleys, then we shall be compelled to seek an outlet in the Napa Valley, where the most active liberality exists in the construction of all needed roads.[110]

By utilizing some carefully chosen exaggerations, Clarke issued an ultimatum to the citizenry of Sonoma County that clearly delineated what the mine owners felt was a strong negotiating position. No one could mistakenly assume that the mines intended in any way to help build the road.

Not long after Clarke's letter reached the board of supervisors, the following notice appeared in the *Flag*: "The Board of Supervisors have ordered an election, to be held on Friday, April 3, to submit the question to the people of Mendocino township and that portion of Washington township which lies east and includes the Healdsburg and Geyser Springs Toll Road, whether they will consent to tax themselves to build the road."[111]

The board had heeded Clarke's warning, but the miners were still not satisfied. Besides deciding that they did not want to contribute to the road's construction, the miners also greatly resented having been placed within the boundaries of the proposed road district. They felt they would be taxed for a road that someone else would build and most benefit from its construction.[112]

A $10,000 bond election was held on April 3, 1874. The total vote was 546, with 373 favoring the road and 173 opposing local taxpayers funding its construction.[113] Later in the month, the Pine Flat Road District was established in order to sell bonds for the road. Robinson and Son was the only bidder for the job and took it with the stipulation that it would be finished within ninety days. This would give Healdsburg at least some of the 1874 summer trade.

Robinson and Son wasted little time in starting construction through Sausal Canyon. It started with a burst of enthusiasm and hoped to be done by July. By mid-May, it had over one hundred Chinese laborers working on the road, using picks and shovels to carve their way up the rugged canyon. In addition to the pick and shovel work, there was considerable amount of rock in the gorge that had to be dislodged by dynamite.[114] Reports from the *Flag* tended to indicate that work was not progressing as rapidly as the

The Robinson & Son building in Healdsburg that served as the Wells Fargo Express office, sending daily shipments to Pine Flat. *Courtesy Healdsburg Museum.*

contractors had hoped, since by mid-June, they had increased the Chinese labor force to well over two hundred men.[115]

In July, the *Flag* regretfully reported "the contractors, Robinson & Son, have found a heavy job in building the road through Sausal canyon" since "there is much more rock that requires blasting." As a result, "they will not finish within the stipulated 90 days."[116] Two Napa County newspaper skeptics had predicted the difficulties earlier, since they perhaps had been able to be a bit more objective. One flatly stated, "The Sonoma route is pronounced almost impracticable for easy road making and light grades."[117] The other had looked at both the surveyor's report and the proposed road and quipped "that instead of it being a 40 gulch road, it should be 100 at least."[118] Nonetheless, Robinson and Son asked for and was granted a time extension as it plodded on with construction.

Road construction dragged on through August as Healdsburg businesses saw their dreams of an active summer trade dwindle. To increase their anxiety, they could read reports similar to the following in their local paper: "Shipping at Calistoga lively—twenty-seven four-horse teams loaded at the depot last Thursday."[119] If they had read Napa County papers, they would have been even more depressed, since they were gleefully detailing the magnitude of the trade into Calistoga. In fairness, one should mention that from four to six teams did leave from Healdsburg for the mines daily, but most were carrying lumber that could not be obtained in Calistoga. Napa County definitely was receiving the bulk of the trade.[120]

It was not until early September that the Sausal Road was completed. The distance from Healdsburg to Pine Flat had been reduced to sixteen miles. The old route had taken nearly four hours with a team of horses, while most agreed that by the Sausal route, the trip up to the mines could be made in two and a half hours and the return in one and a half. With the time and distance reduced, Healdsburg residents had good reason to believe that a fair share of the trade would now be coming into their community. What they had neglected to look into, however, was the matter of freight rates charged by the railroad lines. Much of the freight needed in the mines came from San Francisco. Healdsburg was linked to the Bay City by the San Francisco and Northern Pacific Railroad (SFN&P), which had long been accused of charging exorbitant rates. Complaints about the SFN&P were numerous. "In one instance, the freight on an ordinary two-horse wagon from San Francisco was $18."[121] The Thompson brothers had been staunch supporters of the Sausal Road, but the *Flag* reported in October 1874 that they "get their freight via Calistoga now, because the S.F.N.&P. charge more than the Napa Valley railroad. They tried the Healdsburg route three or four times."[122] Horrified Healdsburg businessmen discovered that they had a better road, but the trade was still going into Napa County. They were then faced with having to fight the railroad. While lamenting this problem, the seasonal rains dampened their hope even further.

As luck would have it, the winter of 1874 was the most severe experienced in the decade. Of immediate concern was the fact that the Sausal Road forded the Russian River at Soda Rock. With the first rains of the season and the resulting high water, Healdsburg was cut off from the mines. The *Flag* had foreseen the problem as early as September and at that time stated, "A bridge is now the battle cry."[123] In October, the *Sonoma Democrat* suggested, "A petition ought to be circulated at once calling on our supervisors to have a bridge built this fall across the Russian River on

Trains entering Healdsburg carried shipments from San Francisco for mines. The new engine in 1872 was named Geyser. *Courtesy Healdsburg Museum.*

the Pine Flat road."[124] However, the enthusiasm of road proponents had waned, and no bridge was forthcoming.

Washouts and slides caused by the winter rains also closed the road. As early as November, an observer reported that he "went a half mile up the road from the mouth of Sausal creek and could not go any farther."[125] In early December, T.W. Hudson had more than a dozen men trying to clear the Sausal Road from the valley end, while Greenville Thompson was reported to have a "force of Chinese" working on the Pine Flat end. As a consequence of the road's closure and in what must have been a humiliating experience to advocates of the Sausal Road, daily stages from Healdsburg to Pine Flat were being forced to "go around by Knight's Valley" and use the Calistoga Road.[126]

The citizens of Healdsburg had undeniably failed to lure the mining trade away from Napa County, but it should be noted that they were not alone in their blind excitement and lack of foresight. Geyserville, situated along the Russian River north of Healdsburg, attempted its own road to Pine Flat

A wagon on Hog's Back on the way to Pine Flat. *Courtesy Library of Congress.*

after witnessing Healdsburg's failure. A Santa Rosa newspaper carried the following account in February 1875:

> *The citizens of Geyserville and vicinity have in circulation a subscription paper to raise funds to build a road from the former place to the mines. Mr. Geo. Petray, who is taking an active part in this commendable enterprise, informs us that the heart of the mining district can be reached in ten miles— that the grades will be less than by any other route—the road when built not liable to slides, so that it may be traveled in winter as well as summer.*[127]

The Geyserville route was, in fact, eventually finished but not before most of the mines had suspended operations.

The city of Santa Rosa was located too far away from the mines to embark on its own road project. Consequently, Santa Rosans were relegated to the secondary role of cheering on their neighboring communities. However, since Santa Rosa was the county seat, it did have one advantage that enabled its citizenry to keep up with the mining excitement. The trials that resulted from the claims of both old and new locators took place in the city of Santa Rosa. Although the trials started soon after the mines opened, newspapers indicated that they did not attract widespread attention until early 1875. The *Sonoma Democrat* of early 1875 carried week after week of full front-page coverage of the trials, printing the entire testimony for three successive issues.[128] An 1877 recollection of the trials reflected the interest that was manifested. The trials "brought to the county seat of Santa Rosa a number of the most distinguished mining lawyers of the Pacific coast, and learned and eloquent arguments were made, which engaged the court for a prolonged session, creating for a time more excitement than was ever before witnessed in any cases in the courts of Sonoma."[129]

The excitement generated over the trials was significant in that it helped illustrate the tragic irony that was connected with Sonoma County's reaction to the mines. Principal factors inhibiting the rapid development of the mines were the lack of capitalistic interest and the fear of litigation and expensive trials. Never realizing this seemingly obvious fact, Santa Rosans took the spectacular trials to mean something entirely different, as illustrated by this quote from the *Santa Rosa Times*: "The interest manifested and the array of talent engaged on both sides in the mining suit now in progress in this city, is sufficient evidence of the value of the mining interest in this county… We confidently look for the investment of millions of dollars of foreign and domestic capital for the development of the mines."[130]

So intense was the interest in the case of *Silas Stone v. Geyser Mining Company* that after the Geyser Mine received a favorable verdict, it was reported that "Santa Rosa 'luxuriated' in an undistinctive general drunk."[131]

In the final analysis of Sonoma County's reaction to its mines, one must conclude that emotion prevailed over reason. The county's citizens, in their exuberance to secure the wealth of the mines, had overlooked numerous facts that in retrospect seem quite obvious. The ultimate irony was that Sonoma County did everything within its power so that it might call the mines its very own, while Napa County did virtually nothing and retained much of the trade in spite of the fact that it was not the "natural outlet."

A Character of the Time

OLD CHIEFTAN

Between 1855 and 1875, the Geysers were second only to Yosemite as a California natural wonder. Just as noteworthy as the steaming fumaroles was the sixteen-mile stagecoach ride up the steep and rugged Mayacamas range of mountains. The ride over the narrow trail was dangerous, but with a character named Clark Foss driving his team of six jet-black horses, the trip was a white-knuckled adventure of terror that no passenger ever forgot.

Clark Foss was born in Maine in about 1819. He came west and settled in Healdsburg in 1859, where he tried to raise hogs but later opened a livery business. For a short time, he was the owner of the Geysers Hotel, but it was his superlative handling of horses that led to his worldwide reputation as a fearless stage driver. Marveling journalists recounted Foss as a huge, cheerful man of over 250 pounds, possessing a thunderous voice that gave him full command of his horses. However, it was as more than a handler of horses that Foss gained notoriety. Some saw him as a wild, whimsical philosopher, wearing a pearl gray Stetson hat and long duster coat, snapping his fourteen-foot whip like the shot of a pistol, hell bent on scaring the daylights out of his passengers.

Known by many as "Old Chieftan," the boisterous and colorful Foss drove his stage out of Healdsburg until he and partners constructed a toll road from Calistoga. The toll road was barely seven feet wide in places and featured steep, unprotected drop-offs into the canyons below. The road went through Pine Flat, and during the 1873–75 mining rush, Pine Flat served as a major stage stop and destination for many passengers.

Left: Clark Foss. *Courtesy Healdsburg Museum.*

Below: Clark Foss's wagon at the Geysers. *Courtesy Library of Congress.*

In about 1865, Foss founded Fossville in Knight's Valley on a level spot just before the ascent into Pine Flat and the Geysers. Fossville was a stage stop with a barn for horses and rooms for travelers. Mrs. Foss fed the visitors. The likes of Ulysses S. Grant, Richard Henry Dana, H.S. Crocker, Mark Hopkins, J. Pierpont Morgan and W.R. Hearst were a few who signed the Fossville guestbook and presumably made the trek through Pine Flat to the Geysers.

Foss carried mail to Pine Flat and undoubtedly hauled potential financiers into town, as well as provided transportation to Chinese and other miners. For Pine Flat folks, he served a practical purpose, and local accounts did not mention Foss's spine-tingling antics. Presumably, to those who lived and worked in Pine Flat, Foss was just one more player in a cast of adventurous characters who were commonplace in the raucous mining town.

In a horrific accident in 1880, Foss plunged his coach and team into a deep canyon, killing one young woman, maiming another for life and seriously injuring four others. Foss himself was hurt so badly he was laid up in bed for some time. Never the same after the accident, "Old Chieftan" handed over the reins of his coach to his son Charles in 1881. Foss died in Calistoga in 1885.[132]

Chapter 5

BUILDING A MINING TOWN

As noted by the *Russian River Flag*, "whenever men and money are found, hamlets, towns or cities spring up."[133] If a few hastily constructed buildings can be used to indicate the beginning of a "hamlet, town or city," the Cinnabar Mining District bore five potential communities during the mining rush. Located near the junction of the Sausal and Geyser Toll Roads was Mercuryville, which boasted several buildings and a post office. Lucky Bill Whitton was said to have been the founder of the town, and the Thompson twins operated a branch dry goods store there. At the mouth of Sausal Creek was Excelsior, which also had a post office and a saloon and hotel run by Martin Bunch. Illinois City was located at the Geyser Springs. A post office and hotel were located there. In the heart of the mining district was Pine Flat, which was the only community that was destined to become a mining town.[134]

Pine Flat's only significant advantage over the communities was its superior location. As noted by the *Flag*:

> *The most eligible site for a town in the new mining region was seen to be the comparatively level strip of country, ten and three-fourths miles northeast of Healdsburg, laid out on Bower's County map as Pine Flat. Level ground in that region is like angels' visits. The strip mentioned may be considered a small plateau, formed by flattening the sharp ridge that divides the head waters of the Little Sulphur (which flows west) from those of the various streams that flow southeasterly and finally empty into the Russian River.*

Pine Flat is from one hundred to three hundred yards wide and about one-fourth of a mile long. Lying in the midst of stately oaks and towering pines, elevated two thousand feet above the sea and enveloped in a pure, cool atmosphere, it is a delightful location.[135]

One may quibble as to how "delightful" Pine Flat's location was, but there was no denying that it was a centrally located twelve-acre plateau in a mountainous area. It was indeed the logical site for a town.

The quicksilver excitement of the 1860s had produced several buildings in Pine Flat, but they were of the crudest variety. The *Sonoma County Journal* of June 15, 1860, had the following account of a fire at "Pine Flat city" with an "immense loss of life!": "Pine Flat City, situated in the mountains near the Geysers, was visited on Sunday night by a terrible conflagration, destroying a canvas covered log building, together with all the inmates, 000,000,000 in number (creeping inhabitants). No insurance—Only one house left standing in the doomed city!"[136] The *Journal's* attempt at humor may have been paltry, but the article did indicate that there was little of importance located in Pine Flat in the 1860s.

In 1872, Pine Flat was a remote strip of land with only an occasional Calistoga stage bound for the Geysers to disturb the serenity. There were no buildings in the Flat, no one lived there and the Calistoga Road was its only link with civilization. The rapid influx of prospectors soon changed everything. However, the first building in the Flat was constructed by a man not connected with mining. In the summer of 1873, Clark Foss, who ran stages to the Geysers, built a stage station at Pine Flat. The authentic town fathers, however, were the Thompson twins, who apparently had decided they were not going to make a fortune as miners. One cannot talk of the early construction of Pine Flat without detailing the activities of Granville and Greenville Thompson.[137]

Starting in the summer of 1873, the Thompsons took up a claim of land that included Pine Flat; laid out a town; began to dispose of lots; built a store, a livery stable and a large hotel; constructed a town reservoir; persuaded the government to open a post office there; and paid for the survey of a township of land. They accomplished all this in little more than six months while taking time to campaign for the Sausal Road and to locate, buy and sell mining claims.[138] The Thompsons were definitely men of action and legitimate fathers of the town of Pine Flat.

All the local newspapers noted the Thompsons' untiring energy and perseverance and considered it a special event when one of the twins descended the mountain to transact business in the valleys. Since the Thompsons were identical twins, the newspapers encountered difficulty in determining which one was actually in town. The *Calistoga Free Press* solved this problem by simply referring to either of them as "Gran-Green Thompson."[139] There is no evidence to suggest that the Thompsons were antisocial, but it appears that they utilized all their time for business transactions and wasted little time on social calls. Their countless ventures substantiate a newspaper account that credited them with "seeing the future as most men see the present."[140]

As noted by the *Calistoga Free Press*, the Thompsons "went into that district when it was a wild waste."[141] Pine Flat was unoccupied government land in early 1873, but the Thompsons quickly perceived its potential and claimed it. To protect their holdings, they paid for a survey of an entire township surrounding the flat. A government surveyor, Gustavius Cox, was employed and completed his survey by mid-June 1873. All the land in the new township could then be preempted or homesteaded, except that which was designated as "mineral land." About one-fourth of the township was so designated.[142]

By the early fall of 1873, the Thompsons were laying out a town and selling lots. In November, the brothers completed their first building, a dry goods store, and Greenville already had departed for San Francisco to "lay in a stock of goods."[143] Before the first of December, Pine Flat had added six houses, Rickman and Sewell had come up from Calistoga and opened a butcher shop and construction had started on a saloon and restaurant.[144] Having completed their store, the Thompsons soon embarked on the construction of a hotel, which was to be completed by spring. In their spare time, they had filed a petition with the government for the opening of a post office in Pine Flat and also had filed an application with the board of supervisors for the establishment of a Pine Flat township, which would include a voting precinct.[145]

By March 1874, Pine Flat had been granted a post office and Granville Thompson appointed postmaster. The post office was named Dodgeville because there was another Pine Flat in the state, but few people ever referred to the town as Dodgeville, and the name was later changed to Pine Flat. By that time, the Thompsons had nearly completed their hotel. They also had commenced building an addition to their store so that they might carry a larger stock of goods. By the first of April, construction began to boom, with three of the mining companies locating offices in the town. In addition, F. Swinney had built a house and opened a blacksmith shop, Sewell and Rickman were adding a slaughterhouse to their butcher shop, J.L. Terry had started a saloon and W.R. Williams was talking of building a hotel. William

C. Graves, a survivor of the Donner Party, was also in the area and talking of various improvements around the Flat, among them another hotel.[146]

By May 1874, construction at Pine Flat was progressing rapidly. The Thompsons began selling lots (50 feet front by 150 feet long) for $200 apiece, and they sold well.[147] Hudson and Ward Lumber Company of Healdsburg shipped considerable amounts of lumber to the town and on May 1 delivered 7,000 feet of lumber over the Geyser Toll Road.[148] Much of the lumber was redwood and logged in western Sonoma County. Heald and Guerne had a lumber business in Healdsburg and also a redwood sawmill in the western part of the county. It was a major supplier of Pine Flat building materials. A reporter in Pine Flat wrote on May 23, 1874, "There are nine regular lumber wagons hauling from Healdsburg to Pine Flat."[149] However, even lumber of this quantity did not satisfy the demand since a mid-May account stated that "three houses are going up and eight or ten more would be under way if lumber could be had."[150] Pine Flat had the first characteristic of a true mining town: hasty construction.

With the coming of the dry season, the Thompsons tackled the problem of supplying water to the community. In the winter and spring, water was

The Heald and Guerne mill near Guerneville provided much of the lumber in the construction of Pine Flat. *Courtesy Healdsburg Museum.*

not a scarce commodity, since a branch of Little Sulphur Creek (sometimes called Cloud Creek) flowed through the middle of town. However, the creek dried up during the summer. With the assistance of Hedge Reynolds and W.R. Williams, the Thompsons proceeded to dam a small spring and stream on the hillside south of town for use as a town reservoir. They procured galvanized pipe and channeled the water into town, running a line down Main Street, the principal thoroughfare in the town. By June 1, 1874, Pine Flat had a municipal water supply.[151]

By October 1874, Pine Flat had not been in existence for one year but nonetheless could boast of thirty-three houses and twenty-four businesses. In addition, many more buildings were under construction. Thomas Hays was putting up a "row of houses 50 feet front and 28 feet deep."[152] Two saloons also were under construction and would add to the already disproportionate number of drinking establishments in the town.[153] As noted on October 23, 1874, in the *St. Helena Star*, "For a mining town of scarcely a year's growth, Pine Flat presents a stranger quite a respectable appearance."[154] In addition, one visitor noted, "Businessmen wear smiling faces, and represent prosperous times."[155]

Largely through the efforts of the Thompsons, construction of the town had been relatively organized. Unlike most mining communities, Pine Flat had been planned in advance—but not much in advance. Only eleven months after the Thompsons had divided the town into lots, an entire community had sprung up. This rapid construction indicates that first-rate materials and carpentry were subordinated for immediate completion. Most of the buildings were constructed hastily from rough redwood planks and in the spring of 1874, "a house belonging to the Lost Ledge Company was blown over."[156] This substantiates a San Francisco correspondent's claim that most of the buildings in the town were "of the primitive style."[157]

The construction of the town reached its climax in early 1875. By that time, there were businesses to serve nearly all the physical needs of the mining community. There were four grocery and dry goods stores, the most notable being owned by the Thompson brothers, who hired their nephew, Eddie Thompson, as proprietor. Others were owned by F. Phillips, Bowen and McGrear and Hyman and Company, whose business was always referred to as the Uncle Sam Store. Hickman and Sewell had sold their butcher shop by this time and returned to Calistoga. However, there were two new meat markets, one operated by Kaiser and Chrisman and the other by Lambert and Congdon. The Pine Flat Market, the name of Lambert and Congdon's business, sported a new wagon and advertised free delivery over

Wagons with high side racks carried lumber for Pine Flat construction. *Courtesy Healdsburg Museum.*

the rugged terrain to the mines. It should be noted that the Pine Flat Market discontinued the delivery service after a disastrous trip to the Missouri Mine. While ascending the treacherous road to the mine, the new wagon, driver, goods and horse "plunged down a declivity and came to a sudden stop, heels uppermost in a deep ravine."[158]

The town also boasted two livery stables: one kept by Ed Kinyon and one by A.A. Brown. The one operated by Brown was rather unique in that he advertised that he supplied mounts or wagons for one-day round-trip excursions to Pine Flat for tourists who arrived in Healdsburg. Brown charged three dollars to rent one of his two-horse wagons for the trip. A fruit store was operated by David Day, while Mrs. Joyce ran the town's bakery. James Hickle was reported to have been running a restaurant, and a merchant tailor shop was operated by M. and A. Jacobs. The town's busy lumberyard was owned by Mathieson and Ferguson, and the Reverend J. Daubenspeck had a brickyard in the town to supply furnace material to the mines.[159]

Russian River Flag, July 1874.

There were also a number of offices in the town. Three of the mining companies had their headquarters in Pine Flat and maintained offices there. Gustavius Cox, the United States mining surveyor whom the Thompsons had hired, had an office and did all the surveying in the district. W.P. Litten opened an office to handle and record mining transactions. The interesting fact concerning Litten's Pine Flat office was that he had reluctantly opened it there after having been run out of Calistoga for allegedly transacting "crooked deals."[160]

In addition to these establishments, there were also a number of other businesses whose proprietors' names have escaped the scrutiny of history. They included two shoe shops, two blacksmith shops, two laundries, a gunsmith and one more lumberyard. A Chinese business operated by Cum Yuk, who had a laundry in Healdsburg, is mentioned once when he left Pine Flat. There were also a number of carpenters and painters who found ample employment in the area but did not have specific places of business. This amounted to twenty-four established and operating businesses exclusive of hotels and saloons. They did an ample job of serving the physical needs of the miners.[161]

Serving the social needs of the area were a sizable number of hotels and saloons. The Thompsons, F. Swinney and later George Reeve built hotels in the town. The first hotel completed was built by the Thompsons, who hired George Reeve of St. Helena to manage it. Completed in early 1874 at a

PIONEER HOTEL

—AT—

PINE FLAT,

SONOMA COUNTY, CALIFORNIA.

GEORGE F. REEVE, •   • Proprietor.

THIS HOUSE, THE PIO-
neer building of Pine Flat,
has recently been reconstructed
and fitted up in first class style by
the Thompson Bros., and will hereafter be kept
in a style commensurate with the wants of that
growing section.

The best Wines, Liquors and Cigars.

The house has Twenty nicely finished and well
ventilated rooms.

Everything conducive to the comfort of guests
will be attended to.

☞ Stages leave Calistoga for Pine Flat, daily.

12

Calistoga Free Press, August 1874.

cost of $5,000, the Thompsons' hotel went by two different names. It was advertised as the Pioneer Hotel in the *Calistoga Free Press*, while the *Russian River Flag* advertised the same place as the Thompson House. Despite the discrepancy in names, both agreed that the building was square (seventy by seventy feet) and contained twenty cottage rooms, most of which faced an open court (twenty-four by twenty-four feet).

F. Swinney built the Pine Flat Hotel, which was said to have been "20 x 36 besides an addition in rear." Swinney had Mrs. M. Davis as proprietor of his small hotel. The third hotel was built and operated by Reeve, who left the Thompsons to compete for their business in early 1875. It was at this time that the Thompsons lured Congdon away from his meat market to replace Reeve as the manager of their hotel. Under Congdon's management, all agreed on calling the establishment the Thompson Hotel. Reeve's Hotel was reputed to have been the biggest and fanciest of the three. It had two stories and a fifty-foot front while being sixty-two feet deep with a dining room of thirty-six by twenty-four feet. Reeve's Hotel could accommodate one hundred guests.[162]

There definitely were three hotels in Pine Flat, but confusion ensues when one tries to determine the exact number of saloons in the town. However, despite the conflicting evidence, one can be sure that the number was great in relation to the population of the town. A newspaper report of October 1874 stated that there were six saloons, with two more under construction.[163] Men named Hays, Howe, Grabiell and Terry are known to have been Pine Flat saloonkeepers. At its peak, Pine Flat had at least eight saloons. However, in December 1874, there was mention of someone "fitting up a hurdy-gurdy house at Pine Flat."[164] There also were other references to "Spanish dance houses" and "bawdy houses," but one cannot be sure if these were included in or in addition to the original drinking establishments. However, one can be sure that houses of prostitution flourished somewhere in or near the town. This probably explains the confusion, since houses of prostitution did not warrant lengthy mention by the newspapers of the day. Pine Flat was undoubtedly like other male-dominated mining towns in having a number of houses of ill repute. They simply did not survive the discriminating eyes of contemporary journalists, who were trying their best to give the town of Pine Flat a degree of respectability.

By the summer of 1875, sixty houses had been built in Pine Flat.[165] As previously noted, many of these were not of a substantial nature, and most were built hastily by individual parties. Not surprisingly, the Thompsons were the only known persons to have attempted to build rentals. A newspaper report of early 1875 stated that the "Thompson Bros. intend to put up five or six tenement houses."[166] The outcome of this particular venture is not known.

A street map of the town indicated that there were eighty lots in the town, and there appear to have been about that many buildings in the town, but there was not necessarily one building per lot. Many of the larger buildings took up more than one lot, and there is evidence to suggest that quite a few

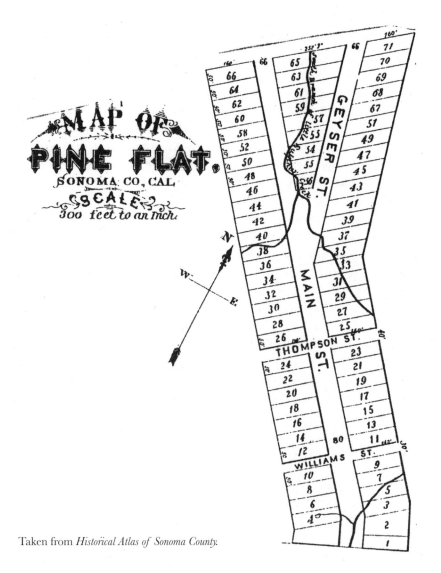

Taken from *Historical Atlas of Sonoma County.*

buildings were located away from the central portion of the town depicted on the map. There were four streets within the boundaries of the town, and one need not look too far to determine from where the names originated. Main Street was simply the main thoroughfare in town, while Thompson Street was a modest tribute to the town's founders. Williams Street was named for W.R. Williams, who owned ten town lots, and Geyser Street was named for the steam springs only a short distance to the north.

Segregated housing prevailed in Pine Flat. The central part of the town was referred to as the "American sector," with most of the businesses located there. A newspaper account noted that Pine Flat could "boast of its two suburban towns as well as any other live town. These two important adjuncts known to all 'camps' as Chinatown and Mexico form part of the place."[167] The two settlements were located at each end of Pine Flat and were composed of tents and shanties. Unable to find lodging anywhere else, the Mexican and Chinese miners were forced to construct their own crude dwellings on the outskirts of the town. Although no eyewitness accounts specific to Pine Flat "camps" exist, they were undoubtedly similar to those in other mining towns: overcrowded, unsanitary enclaves of the most primitive sort. The desire for cheap labor combined with racial prejudice made them inevitable.

Although no photographs of the town are known to exist, one can be sure that, aesthetically, Pine Flat's appearance left much to be desired. The town was hastily constructed to meet the needs of the miners and when building material of even the crudest variety was difficult to obtain. Sanitary conditions were primitive. Consequently, Pine Flat did not resemble the stereotyped western town, with colorful and decorative buildings. Its physical appearance probably could be best described as drab. The suburban "camps" added a ramshackle and odiferous touch. However, one should not assume that its drab appearance and primitive suburbs reflected a demure life in the town.

WILLIAM C. GRAVES

———————————◆►✕◄◆———————————

As a member of the Donner Party, young William C. Graves crossed the Plains and eventually arrived in California in 1847. A native of Mississippi, he lost both parents and one brother on the infamous Donner expedition. He once said he spent his eighteenth birthday at Donner Lake digging a dead horse out of the snow. With Donner Party reports of starvation, cannibalism, brutality and unimaginable hardship, it is safe to say that these events may have affected the young man's mental outlook. Whatever the causes, Graves proved to be a fascinating survivor in more ways than one.

Over six feet, three inches in height and weighing over two hundred pounds, Graves was a presence. Two years after the Donner experience, he returned to the Midwest and guided a party of 49ers from Pittsburgh to California. He eventually settled in Lake and Napa Counties, although "settled" is not an accurate word to describe the man. He worked for a time as a blacksmith in Calistoga. At some point, he fathered several (one report said six) children with a Pomo Indian woman, whom he never married and eventually abandoned. A son from that relationship, also William Graves, became the acknowledged leader of the Pomo Indian tribe in Lake County. In 1873, the elder Graves married Martha Blasdel Cyphers. They were later divorced.

It was not surprising that a man of Graves's makeup was attracted to the excitement that captivated Pine Flat in 1873. Tough, courageous and with an eye for adventure, he was perfectly suited to relish Pine Flat hysteria. Graves staked a mining claim named the Hoosier in the Cinnabar Mining District in 1874. Newspapers reported that he was in and about the town of Pine Flat, talking of building a hotel. Reports suggested he also was interested in a "hurdy-gurdy"

William C. Graves. *Courtesy Marilyn Kramer.*

house and "erecting a commodious dwelling." Graves talked a good line, but there is no record that he built a dwelling or started a business in Pine Flat.

There were no reports of Graves successfully mining any cinnabar ore, although in 1874 he did sell his interest in the Missouri Mine for a reported $5,000. In 1875, it was reported that Graves was running two tunnels to strike a ledge of "float rock very rich in cinnabar." He did, however, stir considerable excitement when he proclaimed that he believed his claim to be rich in gold and silver. Apparently, no one ever saw Graves's gold and silver. He left Pine Flat in 1875 along with the other disgruntled miners and entrepreneurs.

Graves spent most of his remaining years in and around Calistoga, never settling into any specific vocation. In 1891, he was reported to have been prospecting in Lassen County. As a young adult survivor of the Donner Party, he was often asked to recall the events of the tragedy. He gave many informal talks of his experience, with some of his recollections contradicting other versions. In his later years, Graves lived with various sisters and nieces. Some said he did little to make himself welcome, and he was referred to as their "ne'er-do-well" uncle. There were many stories about his eccentric and irascible behavior. One story was that he once split his thumb open, called for needle and thread and stitched it up himself. These stories may hold some truth and may also verify that Graves was an independent free spirit, well suited for the time.

William C. Graves was a restless, adventurous and eccentric risk taker, perhaps the perfect embodiment of Pine Flat's fleeting notoriety. The *Russian River Flag* published an interview with him and called him "emphatically, a pioneer." Who would argue? He died at the Sonoma County Hospital in Santa Rosa in 1907 and was buried in Calistoga. His obituary called him "kind of heart, courageous and firm." A close friend who "was greatly pained to hear of his death" said he was a "splendid shot" and together they had "enjoyed many bear and coyote hunts."

William C. Graves was a true character of the time.[168]

Chapter 6

LIFE IN PINE FLAT

C ontemporary accounts were relatively accurate in describing the physical characteristics of Pine Flat. However, they were less reliable in reporting what it was like to have lived in the town on a daily basis. One can be sure that Pine Flat's populace worked long, hard hours and generally lived a less than romantic life. However, no newspaper or firsthand account depicted the day-to-day drudgery and hard work that built the town and the roads and worked the mines. Most reports noted only the festive and sensational events that transpired. In addition, local journalists did not have an appetite for reporting the sordid details of what they considered were immoral or unsavory activities. Although interesting, a record of only the colorful and respectable aspects presents a somewhat distorted picture of life in the town. Keeping in mind that most of the town's activity consisted of rugged toil for which there were few rewards, one can proceed to examine the available accounts regarding life in Pine Flat.

Although located in a remote region and connected to civilization by narrow wagon roads, Pine Flat was not as isolated as one might imagine. There were numerous stage drivers willing to drive a six-horse team up the winding roads, and most enjoyed the challenge that the roads afforded. One observer described the Calistoga-Pine Flat route as "an ungraded seven-foot shelf on the mountain-side unprotected by wall or fence, with drops of several thousand feet to the valleys and curves so sharp you couldn't have seen anyone coming till too late to avoid a collision."[169] The Healdsburg route through Sausal Canyon had even more sharp curves and was no less treacherous than the Calistoga Road.

The "chief" of all stage drivers, who "handled the ribbons over six jet black horses," was Clark Foss, who drove a daily stage from Calistoga through Pine Flat to the Geysers. With the help of Robert Louis Stevenson, who met Foss in 1880, he became a legend in his own time. In *The Silverado Squatters*, Stevenson devoted a significant portion of the work to describing the flamboyant stage driver who "launches his team with small regard to human life or the doctrine of probabilities."[170] Foss was reported to have weighed a grotesque 250-plus pounds, been gruff in character and possessed a voice that would "wake the dead." Long before Stevenson visited the area, Foss dazzled his passengers with his fifteen-passenger

Excursionists arriving by stage at Geysers Hotel. *Courtesy Library of Congress.*

wagon, which cost $1,150. He had distinguished himself as a daredevil driver without equal.[171]

Besides Foss's Calistoga run, there were three other daily stages into Pine Flat at the height of activity in the region. Once the Sausal Road was completed, two stages arrived daily from Healdsburg and one from Geyserville by way of Mercuryville.[172] Bostwick and Emerson owned and operated one of the Healdsburg stages, while John Martin ran the other. Bostwick and Emerson advertised weekly rates in the *Russian River Flag* as being $1.50 from Healdsburg to Pine Flat, $2.50 to Mercuryville, and $3.00 to the Geysers.[173] Martin's stage ran only as far as Pine Flat, but he boasted that he could make the trip up in two and a half hours and return in about half that time. A.A. Brown advertised a livery stable that provided mounts for a one-day trip from Pine Flat to San Francisco and back.

Although Foss became famous for his spine-tingling antics, one should not assume he was the only daring driver into Pine Flat. Another driver, Robert Safely, attained notoriety by the way he handled a grizzly bear that crossed his track. As an astonished witness noted, "He simply got off his stage, gave him a poke with his whip-handle, accompanying said poke with a left-handed 'God bless you,' and went on his way rejoicing."[174] Other newspaper accounts lamented the frequent accidents that the stages experienced, most of which were caused by two wagons meeting on a blind corner. The most serious accident of the era occurred when Foss's speeding stage plunged into a canyon on a trip from Pine Flat into Calistoga. Seven persons were injured in the accident, with one woman having her "skull fearfully shattered." The woman later died from her injuries.[175]

A good number of the passengers on the stages were curious excursionists hoping to see both curiosities of the area: the Geysers and the town of Pine Flat. Many, such as Tiburcio Parrott, George Beaver, Captain Turner and W.C. Lightner, were San Francisco financiers viewing the mines with an eye on investing capital. Most stayed only for the day, and reports indicate that they spent much time in the bar of the Thompson Hotel, where they often got as "happy as clams at high tide."[176]

In the summer of 1874, Foss's stage carried into Pine Flat Major Henry (sometimes referred to as Harry) Larkins, who was destined to become Pine Flat's most celebrated excursionist. Larkins, described as a dashing, suave and captivating member of the San Francisco press, was in town to map the area and write an article for the *Stock Reporter*. However, while in Pine Flat, he continued to write to his lover, Flora, the young wife of famous British photographer Eadward Muybridge, who served as Leland Stanford's

Russian River Flag, August 1874.

A wagon with excursionists leaving the Russian River Valley for Pine Flat. *Courtesy Healdsburg Museum.*

personal photographer. By intercepting one of Larkins's letters to Flora, Muybridge discovered they "had been on terms of criminal intimacy." Muybridge hunted down Larkins on October 17, 1874. He confronted him at the Yellow Jacket Mine. Witnesses reported Muybridge said, "Good evening, Major, my name is Muybridge and here is the answer to the letter you sent my wife." He then calmly shot Larkins "below the left nipple." The major died an hour later.

Confusion ensued as to whether the crime scene at Yellow Jacket Mine was in Sonoma or Napa County, but the men who witnessed the murder apprehended Muybridge and took him to Napa. Numerous Pine Flat citizens expressed remorse at Larkins's death since he was quite popular in the town. There was some talk of lynching Muybridge, and he finally had a spectacular trial in Napa. His defense made the case that he was temporarily insane due to a recent head injury. The jury did not buy the insanity argument but acquitted him on grounds of justifiable homicide.[177] Shortly thereafter, Flora died of natural causes, and Muybridge had their son committed to an orphanage, convinced the child was not his. Three years later, Muybridge became one of the pioneers of the motion picture industry.

In addition to the excursionists were the Chinese who usually arrived in Healdsburg or Calistoga on the railroad from San Francisco and then took the stage to the mines. In early 1875, the *Russian River Flag* reported that

Tourists at Devil's Kitchen at the Geysers. *Courtesy Library of Congress.*

the Healdsburg–Pine Flat stages were "largely patronized by our Mongolian brethren."[178] The stages were also utilized to carry mail to and from the town. Foss carried a weekly mail run from Calistoga to Pine Flat and the Geysers, and in November 1874, Congress authorized a mail route from Healdsburg to the mining community.[179]

Excepting excursionists, there were few men of sophistication in Pine Flat, while women of any sort were a rarity. Aside from a nucleus of merchants, the town's population was extremely transient. Work in most of the mines was uncertain and seasonal, and this led to a constant flux in the town's population. A report stating that "dress parade officials seem

For much of 1874, daily stages ran from a Healdsburg stage station to Pine Flat. *Courtesy Healdsburg Museum.*

to be unknown in the Pine Flat region" suggested that the populace was a bit rustic in appearance.[180] The transient nature of Pine Flat's inhabitants makes contemporary population statistics nearly meaningless. Estimates of the community's population varied from 100 to 1,500 during the same period. Often writing with more exuberance than fact, local newspapers contributed to the confusion regarding the actual population. To confuse the issue further, Chinese and Mexicans often were not included in population figures. However, from what is known of the number of dwellings in the town and the number of laborers in the mines, it seems safe to say that Pine Flat had well over 200 inhabitants at its peak, while there were probably at least 1,000 more men working in the various mines around the Flat.

The number of women in the town is more easily determined. Because they were a rarity, if not a luxury, numerous newspaper accounts carefully noted the number of females. In February 1874, it was reported, "Five ladies have passed the winter in the town."[181] The largest number of women to have lived in Pine Flat at any one time was reported to have been fifty-five, in contrast to the one thousand or more men in the area.[182] It seems unlikely that many of the fifty-five were housewives; most of them undoubtedly found employment in the saloons and houses of ill repute. Contemporary journalists proved better at counting the women than in describing their daily activities.

Unsubstantiated rumors regarding women and wild nightlife were abundant. For example, it was rumored that the wife of Joaquin Murietta, the famous Mexican bandit, lived in Pine Flat for a time and that she was a notorious character herself. The newspapers' reluctance to detail female activities may have had to do with their conservative valley readership and the desire to portray the mining community as a stable, law-abiding environment. Images of a raucous mining camp might discourage potential excursionists and investors. However, at least one observer found reason for hope, since even though "the female form divine is seen but scarcely on the streets," there was little use of "repining" because "they are a coming, Father Abraham and this town will be the equal of any in the country, even in that respect. Meanwhile, we'll try and be happy in the conviction that bellflowers and tulips, and lilies, and roses will scatter their perfume yet even in Pine Flat."[183]

In early 1875, Pine Flat's male population was jolted when one of the town's better-looking young ladies mysteriously eloped with a young miner. One correspondent half-seriously stated that it would not be wise for the "young thief" to show his face in Pine Flat again. The *Calistoga Free Press* carried this account of the elopement:

> *Pine Flat, beautiful young lady; noble young man without any ducats; young lady don't want any ducats, wants the young man; parents object; a party; young lady the gayest of the gay; young man at the door wrapped in mystery and a huge blanket; two horses; they mount; they swiftly fly; clergyman; yes; yes; five dollars; bliss; whoop la; and that's all we know about it.*[184]

Pine Flat's interest in its women was exceeded only by its devotion to its saloons.

For most of 1874 and early 1875, Pine Flat was a bustling community. The height of the day's activity was about noon when the stages arrived in town.[185] In addition to the stages, there were from fifteen to twenty teams that passed daily through the town on their way to the mines.[186] There also were supplies for Pine Flat's merchants, many of which were sent to the mines as pack mules were reported to "line the street from one end of the town to the other."[187] By the fall of 1874, Wells Fargo and Company had regular freight shipments to Pine Flat, with six to ten teams leaving Healdsburg daily. Not surprisingly, Greenville Thompson was the Wells Fargo agent.[188] All the activity led one Pine Flat resident to report, "Our town is all alive. As I write I look out of the window and can count at least forty men in the street in front of Thompson Brothers' store and hotel. Old Californians say it looks like the days of '49, and strange to say, not an idle man to be found."[189]

There is much evidence to suggest that prices were high in Pine Flat, since the town was a considerable distance from the valleys and the demand for

C.E. Proctor with a wagonload of goods in 1872. Proctor was one of numerous teamsters who hauled goods to Pine Flat. *Courtesy Healdsburg Museum.*

goods definitely exceeded the supply. As noted, lumber was in demand and expensive, costing at least $30.00 per thousand board feet. Hay in Pine Flat was as high as $22.00 a ton, while it could be bought for $12.00 in Calistoga. Pine Flat potatoes cost $2.50 per one hundred pounds, compared to $1.75 in the valleys.[190] Granville Thompson stated in April 1874 that he was doing a "lively business" but could "hardly haul up goods and horse feed fast enough to supply the demand."[191] A correspondent of the *Napa Register* mentioned another amusing indicator of the scarcity of goods: "I have noticed a new way that the youngsters of Dodgeville have in sending their love-letters, viz: writing on an egg and sending it by a small boy. Stationery is rather scarce up there."[192]

The hotels of the town illustrated the degree to which business thrived in 1874. The Thompsons' hotel, of course, was the first to open, and an early 1874 account stated, "Sixty persons took dinner at Thompson's yesterday—a fact which will give you some idea of its importance as a business joint."[193] A visitor to the Thompson Hotel noted that it had "lots of customers, and money is loose, floats carelessly."[194] The Pine Flat Hotel was the second hotel to be completed and was reported to have taken guests even before construction was finished. The mere fact that George Reeve left the Thompsons to build a third hotel suggests that the hotel business was a moneymaking venture. It is safe to assume that these hotels were not patronized by the miners living in the area but rather by curious excursionists, many of whom had little interest in the mines themselves. In 1875 alone, 3,500 visitors signed the hotel register at the Geysers Resort.[195] All of these excursionists traveled by stagecoach through Pine Flat on their way to the Geysers. Many must have stopped and stayed in Pine Flat.

If Pine Flat's hotels indicated that business was good, the town's saloons suggested that a fortune could be made operating a drinking establishment. The saloons catered to the local population, which was quite different from Sonoma County as a whole. In the 1870s, rural Sonoma County was obsessed with the idea that alcohol was indeed evil and that many of the nation's ills could be cured if liquor was made illegal. Consequently, most visitors to Pine Flat sorrowfully noted the overwhelming number of saloons and reported the fact to their local newspapers. The *Sonoma Democrat* once snidely reported that there were "*only* six saloons" in Pine Flat.[196] One report stated that of the first fifteen or sixteen buildings in Pine Flat, seven were saloons.[197] Other accounts lamented the number of "votaries" that zealously wasted their hard-earned wages in the various saloons.[198]

The occasional newspaper references to Pine Flat's saloons were often negative and sarcastic, as saloons were not listed with pride in the manner of dry goods stores or livery stables. The temperance spirit discouraged contemporary newspapers from printing vivid accounts of what actually took place inside their swinging doors. Without any blow-by-blow description of a Saturday night in a Pine Flat saloon, one must be content with piecing together reports of incidents too significant to be smugly overlooked by local newspapers. These reports definitely indicate that Pine Flat's drinking establishments were thriving, rip-snorting enterprises and not places for the faint of heart.

With the mines located short distances from the town, miners had little difficulty getting to the saloons. It was not surprising that the miners spent a good deal of their free time drinking and raising hell since there were few other forms of recreation in or near Pine Flat. They were undeniably a rugged lot, with drinking and gambling certainly higher priorities than bathing and doing laundry. Most of the mines suspended operations on Sunday, and it seems unlikely that it was due to any form of devotion to the Sabbath on the miners' part.[199] The miners used the time off to make Saturday night in Pine Flat anything but a dull and boring experience.

The saloons themselves were flimsy edifices with sawdust on their floors and gaudy pictures on the walls.[200] Pine Flat saloonkeepers performed their duties well, since more than one observer noted that the pockets of the miners were lightened of their surplus cash. Pine Flat whiskey, which a witness said "kills at forty rods," was consumed in great quantity. Besides liquor, there were also cards, dice and billiards to entertain the patrons. For those interested in fisticuffs, "free fights" were said to have been "among the red hot luxuries" of Pine Flat's nightlife. The Mexican miners often resorted to knives to settle their differences, and "blood-letting" was done "free gratis." However, the Mexicans were not the only ones guilty of serious altercations, since a newspaper reported that two mining superintendents partook in a weekend "shooting scrape." It seems safe to assume that anyone who entered a Pine Flat saloon on a weekend ought to have expected to be engaged in various forms of excitement. On Sunday, the saloonkeepers counted their money and the miners licked their wounds in preparation for another week's toil in the mines.[201]

Pine Flat had as many as eight saloons in 1874. *Courtesy Library of Congress.*

There was one notable example of Pine Flat's devotion to its saloons and the lifestyle they promoted. In June 1874, the temperance sentiment of Sonoma County was tested in a local option election. Rural Sonoma County voted overwhelmingly for temperance, but the contest was close in Mendocino Township, of which both Healdsburg and Pine Flat were a part. The *Russian River Flag* quickly noted that the closeness of the contest was not the fault of Healdsburg citizens since they

> *solemnly declared by the large majority one hundred and four, that dram-selling was a public evil and should be stopped. But the will of the voters of this precinct came within thirty-one votes of being defeated by the mining precinct of Pine Flat, which gave seventy-four majority in favor of groggeries, leaving but thirty majority in the township against license.*[202]

One must assume that many of the transient miners did not bother to vote, yet Pine Flat voters had turned out in sufficient numbers to influence the outcome in a manner that clearly supported the town's most successful enterprise.

While there was a Chinatown in the suburbs of Pine Flat and hundreds of other Chinese lived in tents and ramshackle housing at various mines, there is no personal record of their recreational activities. Contemporary newspapers of the 1870s made no mention of Chinese social activities in Pine Flat. With the prevailing anti-Chinese mood and the Chinese custom of keeping to themselves, it was as if the Chinese in Sonoma County were a hardworking, nearly invisible labor force, perhaps best characterized by Anglos as a "necessary evil." However, one should not assume that they did not engage in typical Chinese mining camp activity. Helen Rocca Goss wrote a compelling description of the 1880s Chinese camp at the Sulphur Bank Mine in nearby Lake County. The author detailed how the smoking of opium and gambling (fan-tan game) were routine activities for the Chinese during their free time.[203] There is no reason to believe that the Chinese miners in Sonoma County were much different.

Although there is not enough evidence to suggest that lawlessness prevailed in Pine Flat, one must conclude that the mountain community had more than its share of crime. Several burglary attempts were noted in local newspapers, and an arsonist made quick work of the Uncle Sam Store. According to the *Sonoma Democrat*, the store "was fired yesterday morning at 3:30 A.M. Goods saved in damaged condition; no insurance; the work of an incendiary. Had it not been for the efforts of citizens the whole town would have burned."[204]

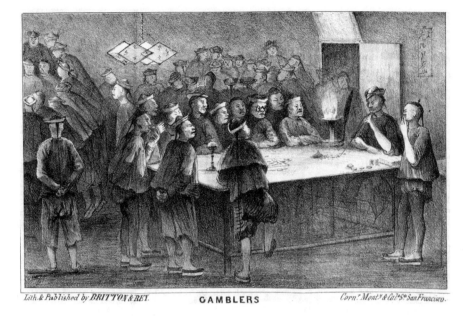

Lith. & Published by *BRITTON & REY*.　　　**GAMBLERS**　　　*Corn.ʳ Mont.ʸ & Cal.ᵃ Sᵗ San Francisco.*

Chinese miners typically spent much of their free time gambling. *Courtesy Library of Congress.*

The wildest section of town was that in which the Mexican miners lived, and while not being saints themselves, the white inhabitants cast a disapproving eye on that part of town. A correspondent sporting a mixture of humor, accuracy and prejudice wrote that "Mexico is doing a rushing business. No dead man for breakfast as yet and no graveyard started. But we hope."[205] The "Spanish dance houses" were allegedly quite rough, and one murder took place when Matthias Salmon killed George Andrado "by shooting him in the temple." A local newspaper carried this account:

> *It seems that three Spaniards, the bar keeper of the house, his brother and another person had been gambling all night and the third had lost all his money. About 5 o'clock in the morning he got up from the table and going behind the bar, helped himself to a drink. The bar keeper interfered and said that it was his business to attend the bar and no one else had a right behind it, and further, that the stranger already owed him six bits. Upon "here's your pay, you s— of a b—," shot the barkeeper through the head. He then took all the money and fled to the hills.[206]*

Reports of attempts to capture Salmon indicate that Pine Flat's law enforcement was primitive. After the shooting, Salmon headed for the hills with Hedge Reynolds and some Mexicans in pursuit. Deputy Sheriff Head went up to Pine Flat two days after the shooting to aid in the capture but returned the next day, having seen or heard nothing of the murderer.[207] With the nearest law officer sixteen miles away, there was little hope of the proper authorities apprehending Salmon, and Reynolds's hastily formed posse had failed. Salmon's escape had been relatively easy. Pine Flat's law enforcement remained several hours away in Healdsburg until Hedge Reynolds was appointed constable in December 1875, but by this time, most of the mines had suspended operations and the population was rapidly declining.[208] It is interesting to note that Reynolds was experienced in vigilante justice, having engaged in a "little pistol practice in a restaurant in Cloverdale," where he exchanged gunfire and wounded a suspected criminal who was later arrested.[209] For most of Pine Flat's existence, vigilante law prevailed, so one might say Reynolds was well suited for the constable job. The following report of a crime suggested vigilante justice was accepted as the norm: "A man was seen stealing a saddle and bridle from the stable, and would have got away with them but for Dick Swift who shot at him twice. The thief dropped the saddle and ran off into the woods. Mr. Swift thought it was a greaser."[210]

The Mexican sector of town was undoubtedly rowdy and probably deserving of condemnation by the white sector, but one should not assume that Pine Flat's "high society" was opposed to festivities with liberal consumptions of liquor. The classic example of the true nature of Pine Flat's leading citizens can be found in examining the Anniversary Ball on November 6, 1874. In the summer of that year, the Thompsons conceived of the idea of having a first-class celebration to commemorate the first year of the town's existence. George Reeve, the proprietor of the Thompson Hotel at the time, made the preparations, and the news of the celebration spread through the neighboring communities. The *St. Helena Star* noted, "Grand preparations are going on, the house is being decorated, turkeys and champagne are going to play a very prominent part and we'll all expect to—well, get stone blind."[211]

Guests arrived from Calistoga, Healdsburg, Geyserville, Middletown and Knight's Valley. It was said to have been a social and financial success with about forty-five couples in attendance. From all reports, the event was a celebration of considerable proportions, since "fun and merriment reigned supreme" and quiet was not restored to the town until "after three days of noise and turmoil."[212] One guest gave this description of the first evening's activity: "After supper, which by all was pronounced A No. 1, it commenced raining, but this only increased the ardor of the merry party within doors and dancing did not cease until 7 o'clock A.M."[213]

Music for the celebration was supplied by Alex Glenn, Al Henly and Joel Harlan, while W.H. Sarles did the calling. They reportedly "discoursed sweet music" while "the wine flowed freely."[214] Because of the rain, most of the dancers had to lay over, and consequently, they improvised another dance that lasted a good part of Sunday.[215] Pine Flat's Anniversary Ball may have attracted a relatively sophisticated crowd from the valley communities, but they certainly adopted the miner's spirit once they arrived in the mountains. One can only speculate as to how many of the revelers were supporters of the temperance movement that was popular in their home communities.

There were two other noteworthy celebrations during Pine Flat's brief existence. In August 1874, the Thompsons held a sizable "social party" in

Visitors dressed for a "mountain event," leaving Calistoga for Pine Flat and the Geysers. *Courtesy Healdsburg Museum.*

honor of their visiting nieces. The event was held in the Pine Flat Hotel and well attended.[216] In April 1875, George Reeve held a celebration on the opening of his new hotel. The opening ball had sixty-five couples in attendance, with two stage loads of guests arriving from Healdsburg and two wagonloads from Geyserville. The guests enjoyed an "Epicurian feast" and were reported to have "imbibed freely" since "free drinks were the order of the day."[217]

Led by the Thompsons, Pine Flat's established citizenry, along with their valley cohorts, found more acceptable forms of entertainment than that of the miners, but there was no evidence to suggest that they were their moral superiors—at least not while they were in the mountains. The Thompsons had energetically considered the physical and social needs of Pine Flat, but they wasted little effort on the religious and educational needs. One resident of the town noted that the "moral atmosphere needs purifying," but there were few "purifiers" ready to step forward.[218] In all fairness, one might note that a moralist in Pine Flat probably would have had a sparse following. The *St. Helena Star* reported that the "absence of a church seems not to be felt very severely."[219] Even though no church existed, the town's brick maker, Reverend J. Daubenspeck, gave Sunday sermons, but they were poorly attended. The good reverend was not a typical clergyman. Before going to Pine Flat, Daubenspeck preached open-air sermons in front of the Union Hotel in Healdsburg for "persons who never go to church."[220]

There was no schoolhouse in Pine Flat during the 1870s, but it is interesting to note that there was talk of building one, even though there were few school-age children in the area. One correspondent was skeptical of the motives behind schoolhouse talk and stated that he believed that it was "only a spasmodic effort at righteousness, or what is more probable, the young men and old bachelors indulged in visions of a good-looking schoolmarm coming up here and settling amongst us."[221]

As noted, righteousness appeared only sporadically in Pine Flat, and the moral atmosphere undoubtedly did need "purifying," but it should be noted that this was not atypical of youthful mining towns. The town emerged spontaneously to serve the immediate needs of the miners, a roguish collection of predominantly young males in search of wealth and adventure. Schoolhouses, churches and law enforcement agencies were of much less immediate concern than were livery stables, dry goods stores and saloons. However, they would have eventually found their way into Pine Flat. Hedge Reynolds's appointment as constable and Reverend Daubenspeck's sermons were indications that the town was mellowing with age. However, Pine Flat was denied respectability by a premature demise.

GRAN-GREEN THOMPSON

Granville and Greenville Thompson were identical (some would say interchangeable) twins born in Missouri in 1844. As toddlers, the twins and six siblings, along with their parents, Micajah and Sarah Thompson, moved to Arkansas. The 1860 U.S. census showed the family left Arkansas and moved back to Missouri for a time and then on to Napa County, California. A later newspaper article added that the twins had also been to Louisiana before landing in Napa.

As teenagers, the Thompson twins proved they had the potential to be movers, if not shakers. What is more difficult to prove was which of the identical twins actually did what. Although those who knew them called them Granville and Greenville, census records, voting rolls and newspaper accounts variously identify the twins as Samuel, Joseph, James, Daniel, S.J., J.G., and G.J., often with Granville or Greenville as middle names. While confusing and often contradictory, their official names appear to have been James Greenville and Samuel Granville. One can speculate if the twins knowingly contributed to the name confusion, but it does help explain why eventually they often both became known as Gran-Green.

Census records indicated that in 1870, the twins' parents and several siblings were in Napa County, living near Monticello. Since Micajah Thompson had always been a farmer, it is safe to say the Thompsons were engaged in some form of agricultural activity. Records list Granville and Greenville as living near the family, although there were reports that Greenville built a hotel in Monticello and Granville worked on the Calistoga

Road to the Geysers. However, from 1860 to 1872, there is little public record of what the Thompson twins were actually up to in Napa County, although there was evidence that they were not passive young men. The May 16, 1868 *Napa Register* reported, "A serious cutting affray took place at Monticello, Berryessa Valley, on Wednesday last, in which a man named Aeckl was badly injured by one Greene Thompson."

The same paper later reported another noteworthy event involving the Thompsons that transpired on May 1, 1872. While walking on the main street of Monticello, the twins encountered the three See brothers, with whom they had a lingering feud. They exchanged angry words, and Granville Thompson and John See agreed to "fight it out." They clenched, and after scuffling a short time, Granville drew a dirk knife and succeeded in inflicting two slight wounds on See's shoulder. Greenville then "rushed up and drew a pistol and shot See." One of the other See brothers then drew his pistol, but it was taken from him, and no more shots were fired. The coroner's report stated that the shot was fatal and the "deceased expired in a few seconds." In a later account, the event was called a murder. Granville stabbed a man with a knife and Greenville shot him dead, but for whatever reason, no arrests were made and there was no follow-up mention of the incident in local newspapers.

One might assume that there remained some bad blood between the surviving See boys and the Thompson twins and that perhaps it was no coincidence that the brothers soon left Napa County. Benefiting from a far west culture that had little concern for one's past, the Thompsons apparently put the "murder" behind them and moved on. While working on the Calistoga Road in 1867, Granville had noticed red cinnabar ore in the nearby hills, and in 1872, the price of quicksilver was on the rise. The twins had their eyes wide open and were probably eager for a change of scenery— and as the surviving See brothers may have attested, the Thompsons were risk takers who needed to be taken seriously.

In 1873, the Thompsons seemingly came out of nowhere to accomplish in two years a list of feats that was nothing short of amazing. They staked mining claims in the Cinnabar Mining District, did some mining, purchased land in Pine Flat, laid out a town, sold lots, developed a water system, built a hotel, opened a dry goods store (with a branch in Mercuryville), lobbied the board of supervisors for Pine Flat Road and were the unrivaled leaders of the area. In addition, one or the other served as Wells Fargo agents, postmaster and notary public; sat on the Pine Flat Road Committee; was appointed overseer of Pine Flat Road; supervised the building of a road from Pine Flat to the mines; and hosted a number of memorable social events.

Barely thirty years of age, these bachelors were miners, entrepreneurs, public servants, civic leaders, road builders and visionaries while accumulating significant wealth from the sale of mining claims and town lots. Local journalists were fascinated with their drive, marveled at their achievements and mentioned them in nearly every issue but rarely quoted them. Articles reflected respect, if not fondness, saying the twins were always in a hurry with little time for conversation. Despite their social quirks, these determined dynamos were undeniably the most influential power brokers in the Cinnabar Mining District.

After being center stage during Pine Flat's heyday, the Thompsons quickly returned to obscurity when quicksilver mining died in 1875. They owned property near Pine Flat and quietly stayed a few years, living in abandoned buildings and running sheep in the nearby hills. An 1875 news report stated J.G. Thompson was seriously injured in an accident. Not much else about them was reported until the following story on March 18, 1878, under the title "Murder at Pine Flat":

> The Calistoga correspondent of the Napa Register writes in that paper the following particulars of the double murder: "The Geyser mail carrier, who arrived at this place Wednesday night, brought the news of the murder of the Thompson brothers, Granville and Green, who lived at Pine Flat. Our informant says that while sitting in the house, after night, they were fired upon through a window, one being killed instantly and the other dying Monday morning—the shooting having been done on Sunday night. Today the story is corroborated by another party, who says they were at Pine Flat the night the murder was committed. The following circumstances are supposed to have been the cause. Green and Granville Thompson (twin brothers) had a fight with a man by the name of See, in Pope Valley, some years ago, which resulted in the death of the latter—the trouble originating through some land affair. It is said the brother of Mr. See vowed vengeance. After years of lawsuits, the Thompsons escaped the law, and have since acquired a fortune at Pine Flat, through mining enterprises and sheep raising, having at present 12,000 head of sheep on their ranch. The story goes that See (who, it is supposed, did the shooting) was seen at the Geysers several days previous to the killing of the Thompsons. As they were

at Santa Rosa, it was necessary for him to wait their arrival home—Sunday night being the first night they had spent at home for over a week. The man who did the shooting, it is said, lives at Eugene City, Oregon, and has taken his trip to California on purpose to wreak vengeance on those who caused his brother's death. Parties who are acquainted with the Thompsons say that they have long expected an attack on their lives, and in all cases went heavily armed. But on this occasion they had no chance for defense, as the shooting was done from the outside of the house, and with a cool and deliberate aim."

The above story was quickly repudiated by the *Russian River Flag* when it reported, "The *Napa Register* has been rudely hoaxed into publishing in all seriousness the 'Murder of the Thompson boys at Pine Flat.'" It went on to say, "One of the brothers was in Healdsburg yesterday and stated that they were very happy to notify their friends that no obituaries were in order yet." While the twins may have been "happy" to notify friends that their reported deaths were a hoax, the mere fact that a scurrilous hoax was perpetuated would lead one to believe that there were folks who wished, in fact, that they were dead. One can reasonably assume the surviving See brothers (as well as other Napa County folks) harbored some serious ill will and that the Thompsons well may have expected an attack on their lives. It might also explain why the Thompsons adopted a low-key profile and may have, in all cases, been heavily armed.

An 1879 article reported the brothers had lost forty-one head of sheep that fell into an abandoned mineshaft and suffocated. For whatever reasons, they had forsaken their hectic and visionary lifestyle for that of domesticated settlers. After a major Pine Flat fire in 1880, a news article reported the fire was of an "incendiary nature," and the brothers collected $2,000 of insurance money for the loss of their abandoned hotel and $200 for their vacant store. Since the Thompsons were conveniently away from Pine Flat when the fire started, one might speculate as to whether they had devised a clever way of cutting their losses. Nonetheless, an 1890 recollection bluntly stated they went from being "considered wealthy" in Pine Flat to being "identified with a drab life."

Records indicate that both were married by age thirty-six, Greenville to Virginia (Jennie) Melton and Granville to Amelia Richardson. Both women were nearly half their husbands' age, and each soon gave birth to a son and a daughter. Greenville's infant daughter died in 1882. The 1880 census showed both brothers as farmers living in Pine Flat. Granville's household included his wife, two children, mother-in-law and a sixteen-year-old Chinese servant. Greenville housed his wife, surviving son, a twenty-two-year-old nephew and an unrelated eleven-year-old girl. Sometime after the fire ravaged much of Pine Flat, the families moved, Granville to Mendocino County and then back to Sonoma County and Greenville to Lake County. From there, the Thompson twins disappeared from historical view, showing up only in occasional voting or census records. In 1896, Greenville was registered to vote in Lake County. The 1900 census listed Jennie (Virginia) Thompson as head of a household with six children. There was no mention of Greenville, who died in 1904. In the 1910 census, a widowed Granville was living with a daughter and her family in Santa Rosa.

During the 1870s, life in Pine Flat was synonymous with the eclectic adventures of Granville and Greenville Thompson. They were part of everything that happened there, and like the town itself, their importance and notoriety vanished with lightning speed. As young risk takers, they typified the hardworking optimists who descended on the Golden State during the westward movement. Their checkered backgrounds, violent encounters and eventual adoption of sedentary lifestyles also typified many early Californians.

Sadly, there are few firsthand accounts detailing the personal qualities of Granville and Greenville Thompson. One is left with a nagging curiosity as to what these fascinating men were like in their daily lives. They were bright, ambitious, persuasive and, most certainly, impulsive, but the actual personality traits that shaped the aspirations of these movers and shakers are elusive. For two years, they displayed invincible determination, but one can only speculate as to what they planned if the quicksilver price had not plunged in 1875.

To this day, the Thompson twins remain shrouded in mystery. Journalists of the time often tossed about flowery platitudes regarding Pine Flat's citizens. The brothers hired their sixteen-year-old nephew, Eddie Thompson, to manage their dry goods store in 1874. Eddie was glorified in a *Calistoga Free Press* article as a lad of "honest and courteous demeanor" who "made a host of warm friends, and helped to build up an extensive trade enjoyed by his uncles—the Thompson Bros." Unlike Eddie, Granville and Greenville were mentioned often in the press but never in a personal or flattering manner. The absence of characteristic complimentary prose regarding the twins' personality is interesting—and perhaps revealing.

No known photographs exist that might give one a chance to look them in the eye and maybe get a glimpse into their personas. If not charming and personable, how were they so persuasive and successful? Were they intimidators prone to violence? Were they simply driven and too busy for niceties? Were they stone-faced serious? Aloof? Devious? Ruthless? Who wanted them dead? Why did they suddenly disappear from the limelight after two years of relentless activity? Why is there no record of their ever again displaying the shrewd business acumen or civic power that created the lively boomtown of Pine Flat? And the confusing trail of name changes—did it have anything to do with that suspicious shooting in Napa County? They were most certainly a duo to be reckoned with.[222]

Chapter 7

THE RAPID DEMISE

In the year following the prospecting rush, Pine Flat was the scene of extensive mining development, the terminus for a new road and the location of a bustling community. By the fall of 1874, the mines had begun to produce significant quantities of quicksilver, and the growing town of Pine Flat industriously served the needs of the miners. On the surface, lasting prosperity seemed imminent. However, there was one factor that could eliminate the activity as rapidly as it had started it: the price of quicksilver. In early 1875, the price had fallen from its high of $1.65 a pound to $0.90, and by April, it reached $0.60 a pound.[223]

None of the infant mines that were attempting to commence operations could withstand the price drop. With the initial cost of setting up reduction works, owners could not hope to mine profitably with the price at sixty cents. The established mines could continue only by cutting costs and reducing the ore they had on hand. By summer 1875, most had laid off much of their labor force and were working slowly in the hope that the price would rise. Blaming a devious "quicksilver ring" that forced the price drop, a local newspaper lamented the plight of many miners: "The men that worked the hardest fared the worst. The miner came back from his tunnel to learn that he was no longer needed, and that he must wait an indefinite period for his wages unless he would take his pay in unsaleable mining stock."[224] However, the trend did not reverse, and by the end of the year, the price had dropped to fifty cents a pound.[225]

By the fall of 1875, it should have been obvious that Pine Flat and the Cinnabar Mining District mines were doomed unless there was a sudden reversal in the price trend. However, even as the price was steadily dropping and people were leaving the area in droves, the valley optimists remained undaunted. In what had to be one of its greater understatements, the *Russian River Flag* announced in November 1875 that "the depression in quicksilver has put a damper on Pine Flat."[226] At about the same time that the price hit fifty cents, the *Calistoga Free Press* somehow found reason to write that the "town of Pine Flat is enlivening, and has undoubtedly seen its worst days."[227] The *Sonoma Democrat* chose to ignore the grave realities and look on the brighter side of things. It stated that because of the depression, "the floating population has departed elsewhere, the field no longer affording place for that unhealthy class."[228] It failed to note, however, that the "unhealthy class" was the labor force of the mines, without which there would be no mining.

The miners could not afford to share the journalistic optimism. The Rattlesnake Mine was the first to suspend operations completely when the price dropped. It closed during the summer of 1875, while the others attempted to keep busy until the rainy season.[229] The timely closure of the Rattlesnake seemed to validate an earlier assertion by the *Russian River Flag* that its superintendent, B.F. DeNoon, did indeed possess a keen "business-like" acumen. With the coming of winter, all the mines closed. Only the Oakland reopened in the spring of 1876. The Oakland staggered through 1876 and managed to produce until 1879, when it also closed.[230]

Not unlike other mining towns, Pine Flat was totally dependent on the mines. Stripped of its only economic base, Pine Flat was destined to disappear as rapidly as it had emerged. After a year and a half of bustling excitement, 1876 proved to be very sedate. Few people lived in the Flat, and those who remained did so only in the hope that the price would somehow miraculously rise. The only noteworthy event of 1876 was the fire that consumed three vacant buildings. In August, an observer noted, "Pine Flat is as dead as Goldsmith's 'Deserted Village.'"[231] By early September, mail service with Healdsburg was discontinued and the post office was closed. The last business to leave the town was owned by L.A. Phillips, who had opened one of the original stores in the town. A newspaper account noted that "Mr. Phillips was one of the five merchants who commenced at the Flat, and is the last one to close up."[232] Pine Flat was indeed dead.

An interesting and somewhat ironic aspect of Pine Flat's rapid rise and sudden demise was that both were out of the control of those affected. Unlike other deserted mining towns, Pine Flat had not died because the mines had been worked out. In 1876, relatively little of the quicksilver had been touched. Pine Flat was created by a quirk in the intricate process of supply and demand and perished for the same reason. Local parties had nothing to do with the price rise and had very little to do with its drop. In fact, if the newspapers were valid indicators, local citizens only vaguely understood the process at all. Perhaps the miners would have better understood their plight had they flooded the market with quicksilver, but in two years, they had produced scarcely enough to be noticed on the world market. However, despite their failings, two simple factors contributed to the fall in price: the increase in the supply and the decrease in demand.

The Comstock mines of Nevada had been the principal user of quicksilver, but by 1876, they had begun their final decline.[233] In addition, the initial price had spurred on the established quicksilver interests, and most had greatly increased production. Promise of profits also had enhanced the development of numerous new mines, many of which were more successful than those in the Cinnabar Mining District. In California alone, production rose from 27,756 flasks in 1874 to 75,074 in 1876.[234] California's production (see chart) shot upward shortly after the price increase, and the result was the price decline. Advertisements listing used mining equipment were common in industry newspapers.

The mines of the Cinnabar District could not operate profitably in the 1860s with the price at or near sixty cents a pound and also could not do so in the 1870s. There was definitely cinnabar ore in the Pine Flat area, but it was not rich enough to allow the mines to prosper if the price was low. In addition, the miners of Sonoma County had proven that they did not rank with the best in terms of efficiency. One can only speculate as to whether the mines could have survived if they had sufficient capital and able management. However, in light of later attempts to reopen the mines, it seems unlikely that even an efficient operation could have mined the low-grade ore profitably once the price fell. The ore in the area was of marginal value, and this combined with quicksilver's fluctuating price made the mines and the accompanying town of Pine Flat ephemeral enterprises.

Indicative of the topsy-turvy nature of Pine Flat's existence were the adventures of the Thompson twins. There was much evidence to suggest that the brothers were shrewd risk takers, as well as energetic businessmen. They entered the region as speculators and made the most of the time they spent there. Unlike the local journalists, the Thompsons were able to keep the excitement in proper perspective. They came into the area penniless in 1872. By locating some claims and selling them when few others were willing to sell, they immediately had access to $30,000, a sizable sum of money during the recession of the 1870s. They used a good part of their cash to build Pine Flat into a bustling community, in which they operated at least three of the businesses that charged high prices. They also sold a number of town lots, which had an assessed valuation of $40 each, at $200 apiece.[235] While the town flourished, the Thompsons were its premier citizens, but when the town declined, they perceived the gravity of the situation as readily as they had the potential and quickly cut their losses.

In early 1875, Pine Flat was the busiest community in northern Sonoma County, but one year later, it was a ghost town. Nearly all the mine owners lost whatever they had invested, and the sum often ran into thousands of dollars. The citizens of Healdsburg and the road district were out $10,000 for a road that never brought much money into the town and was virtually useless once the mines closed. Soon after the town was abandoned, Dio Lewis, an eccentric feminist and temperance leader, passed through Pine Flat on a visit to the Geysers. In an 1881 book, Lewis wrote that Pine Flat was a "played-out quicksilver village" and claimed to "know one case where a man sold a good quicksilver mine for a hunting dog, and it turned out the dog wouldn't hunt."[236] A visitor to the Flat in 1879, while using questionable population figures, described the situation as follows:

> Pine Flat, Sonoma County, is a good example of a deserted mining town. Five or six years ago, when this part of the country was all excitement on account of the quicksilver discoveries, Pine Flat was a lively town of three or four thousand inhabitants. Six or eight hundred men were to be seen in the streets, the hotels could not accommodate all who came and business of all kinds was just "booming." Houses were put up in a day, canvas tents stood upon every available spot, a system of pipes supplied water to the town, and the wave of prosperity threatened to swamp everybody with riches. But a change came o'er the spirit of their dreams.[237]

It was as a ghost town that it next attracted attention. In 1880, a good part of the town was still standing, but the buildings were boarded up and only a handful of people lived in the area.[238] It was at this time that Robert Louis Stevenson was in San Francisco with his new bride, the former Fanny Osbourne. Plagued by consumption and lacking funds, Stevenson conceived of the idea of heading into the area near Mount St. Helena. Through some friends and Clark Foss, the Stevensons heard that in that once prosperous mining area, "the mines had petered out; and the army of miners had departed, and left this quarter of the world to the rattlesnakes and deer and grizzlies."[239] As noted in Stevenson's *The Silverado Squatters*, "It was with an eye on one of these deserted places, Pine Flat, on the Geysers road, that we had come first to Calistoga," since "there is something singularly enticing in the idea of going, rent-free, into a ready-made house."[240]

In her book *Our Mountain Hermitage*, which deals with Stevenson at Silverado, Anne Roller Issler states:

Robert Louis Stevenson in 1880, photographed in San Francisco shortly before visiting Pine Flat. *Courtesy Robert Louis Stevenson Museum.*

> *This phantom place might indeed have provided a free roof over the head of Robert Louis Stevenson and the wife he was unable to support. But they must have food as well as a roof. Clark Foss, stopping at Pine Flat to change horses, could have delivered meat and groceries, but in those days, before refrigeration, he could not have been relied on for unsoured milk. Humorously, the city-bred invalid, for whom fresh milk was essential, considered buying a cow. But this would have involved taking a field of grass and a milk maid.*[241]

Reluctantly, Stevenson was forced to give up on Pine Flat, and as he noted, "It is really disheartening how we depend on other people in this life."[242]

Thwarted in their effort to live at Pine Flat, the Stevensons finally settled at Silverado, a smaller ghost town located on the slopes of Mount St. Helena. At Silverado, they found it easier to procure the needed supplies. It was there that Stevenson wrote *The Silverado Squatters*, his entertaining and somewhat romanticized version of life in the Mayacamas. The fact that he and Fanny stayed only two months may have been partially due to events about which Stevenson wrote in his journal but chose to not include in his book. Examples included numerous encounters with rattlesnakes and Fanny's unfortunate bout with poison oak. Nonetheless, it is possible that Stevenson's need for fresh milk kept Pine Flat from being immortalized by the pen of the famous author. Stevenson's change of mind destined Pine Flat for oblivion. But as noted by Issler, it was probably just as well that the Stevensons moved on to Silverado, because on a Sunday night in July 1880, a fire swept Pine Flat. The Thompsons collected their insurance money, but the Stevensons "might

A woodblock engraving by visiting friend Joseph D. Strong of Stevenson and his wife, Fanny, in their Silverado cabin. *Courtesy Robert Louis Stevenson Museum.*

have lost what little they had in the conflagration" and "might have even lost their lives."[243]

Stevenson's short stay in the Mayacamas area did provide a revealing glimpse at the diversity of people who were drawn to the region. Wealthy excursionists and sophisticated curiosity seekers were in stark contrast with the rough-and-tumble folks who had been afflicted with quicksilver fever. In *The Silverado Squatters*, Stevenson wrote that he met a class of people he called "Poor Whites or Low-downers" who were "rustically ignorant." He wondered if "this is a form of degeneracy common to all back-woodsmen." Bill Spiers, a teamster who sometimes drove a stagecoach, provided an opposing view. Spiers recalled that he transported Stevenson and was alleged to have said about him, "Fact is, I thought him kind of a fool, livin' in that old shack awritin' books!...I didn't think he was hardly as smart as I which had got my diplomy back in Kintucky when I was seven and a half."[244]

After the 1880 fire, only seven buildings remained in Pine Flat.[245] The remaining seven probably burned at a later date or were torn down for their lumber by later visitors to Pine Flat. If any of the original buildings were left in Pine Flat by 1904, they perished in a large wildfire that swept through the area in the summer of that year.[246] A schoolhouse and saloon were built on the site of the original town sometime after 1904. They served the social needs of the few miners who lingered in the area and the educational needs of the small number of children who lived near the Flat. The school was eventually closed, and the area became part of the Geyserville school district. The schoolhouse and saloon were eventually consumed by fire.

As for the mines, they were worked intermittently around the turn of the century, but the price never regained a level to make it truly profitable. The period from 1889 to 1907 was the most productive, but the production figures did not match those of the 1870s. It was not until a demand for the wartime use of quicksilver that the price rose high enough to revitalize an intense interest in the Cinnabar Mining District mines. During the First World War, the *Geyserville Gazette* carried the following description of the activity around Pine Flat:

> *The U.S. Development Mines Co., operating the quicksilver properties in the Pine Flat district, is building a new furnace of 50 tons daily capacity. On the Rattlesnake Mine in the same district, three retorts are now running and nine others are to be constructed at once. A bridge is being constructed over Sausal Creek at the foot of the grade leading to Pine Flat. The entire road is to be put in first-class condition at once*

An unidentified mine with tunnel entrance near the Geysers. *Courtesy Healdsburg Museum.*

in order to provide the means of prompt and easy transportation to and from the mines. The various enterprises now under way will give steady employment to forty or fifty men.[247]

One can imagine elderly skeptics recalling similar enthusiasm in the 1870s.

Even though the mines were working again and the Sausal Road was open, it was not like the "days of '74." Many of the old mines had been consolidated into several holdings, and relatively few men were working in the mines. The community at Pine Flat consisted of only a few buildings. The old road to Calistoga was no longer traveled, and whatever trade there was was sent unceremoniously into Healdsburg over the improved Sausal Road.

The mines were again worked in the 1930s and during World War II, with the Culver-Baer, Cloverdale and Socrates being the most productive. The Culver-Baer Mining Company consolidated the old Oakland, Geyser and Missouri Mines, and by 1946, it had produced 12,000 flasks of quicksilver. The Cloverdale Mine also consolidated a few abandoned claims and produced 17,000 flasks. The Socrates, which had previously been known as the Pioneer, produced 5,500 flasks up to the end of World War II. In addition, there was also some work done on the Contack Mine, which had not been worked during the 1870s. The significant factor about the sporadic workings in the twentieth century was that they confirmed the fact that cinnabar was present in the area, but the ore was not rich enough to warrant reduction on a steady basis.[248]

Today, nearly a century and a half after Pine Flat's heyday, there is little to remind the casual visitor that anything out of the ordinary ever transpired on the chemise-covered ridges that compose the Cinnabar Mining District. Most of the original mine sites have been obscured by nature, with chemise brush covering the caved-in tunnels and forest fires erasing any trace of the original buildings. The most newsworthy events in the area since the 1870s have had to do with the development of geothermal power.

There were several unsuccessful attempts in the early twentieth century to capture steam power in the Mayacamas region. In the late 1950s, Pacific Gas and Electric (PG&E) drilled steam wells seven to eight thousand feet into the earth. In 1960, PG&E completed the first commercial geothermal electric unit in the Western Hemisphere. In addition, many other companies showed interest in developing steam power. PG&E produced significant geothermal power for over a decade, but the wells eventually suffered from diminished pressure loss in the steam reservoir. In 1989, Calpine Corporation entered the picture and led efforts to study water injection to create steam

Above: The Geysers before being capped for geothermal power. *Courtesy Healdsburg Museum.*

Left: Pacific Gas and Electric's first geothermal installation in 1959. *Courtesy Healdsburg Museum.*

to sustain reservoir pressure. In 1995, Calpine and Lake County Sanitation agreed to construct a twenty-nine-mile pipeline to the Geysers to deliver treated wastewater. Between 1998 and 2003, a forty-one-mile pipeline was built from Santa Rosa up Pine Flat Road to the Geysers. Today, Calpine utilizes Santa Rosa and Lake County wastewater pumped to the Geysers to generate geothermal power. Calpine is a multinational corporation and the largest supplier of geothermal power in the country. It is estimated that Calpine's fifteen power plants at the Geysers provide nearly 60 percent of the average power needs in the region from the Golden Gate Bridge to the Oregon border.

In 2012, the actual site of the town of Pine Flat and the surrounding area was acquired by Audubon Canyon Ranch and became part of the 1,620-acre Mayacamas Preserve. In 2013, Audubon Canyon Ranch assumed ownership of the 1,750-acre adjoining Modini ranch property. The resulting 3,370-acre Mayacamas Modini Preserve is now a protected area open to the public for nature education, conservation science and habitat protection and restoration. Audubon Canyon Ranch staff and volunteers operate an extensive stewardship program.

Audubon Canyon Ranch sign on Pine Flat Road. *Courtesy Mickey Bitsko Enterprises.*

Culver-Baer Mine was worked in the 1970s. This photo was taken from Geyser Road, with Sulphur Creek and the Geysers over the first ridge. *Courtesy Patrick Dirden Photography.*

Although of little interest, the undisputed outlet for the area is now Healdsburg. The Calistoga Road long has been closed to public travel. The Sausal Road is maintained by the county and is now known as Pine Flat Road. The road has been improved but follows the original route and is still a narrow, crooked thoroughfare that winds its way through Sausal Canyon. It crosses numerous gulches and passes scarred rocks, which are silent tributes to the muscle of "our Mongolian brethren." It follows the last treacherous leg of the original Sausal Road, which the stage drivers referred to as Cape Horn. It is from the Cape Horn section of the road that one can look back toward Healdsburg, where vineyards have replaced the "fields of waving grain." Looking in the other direction, one can see Knight's Valley and imagine Clark Foss driving a six-horse team and coach at full speed "with small regard to human life or the doctrine of probabilities." Today, pickup trucks cautiously travel the improved Pine Flat Road. Occasional bicycle enthusiasts with high-speed mountain descents rival Foss and his daredevil contemporaries.

Although ravaged by recent fires, Pine Flat itself still supports steely oaks and towering pines for which it was named. The small branch of Little Sulphur Creek still meanders through their midst. Today, the large clearing is covered with underbrush but still appears to be the logical site for a town. However, the only hint that there ever was a settlement there are the occasional depressions in the earth that were once the foundations of buildings that harbored excitement that is hard to imagine ever existed in this serene enclave. If one searches for it, the spring can be found on the hillside "south of town" that served as a municipal water supply. Nestled in thick Manzanita north of town can be found the Pine Flat cemetery, which retains the remains of some long-departed citizens. Pine Flat Road runs almost parallel and slightly west of what was once Main Street. Yet from Pine Flat Road, there is only an occasional "wildlife sanctuary" or "no hunting" sign to remind visitors that civilization ever intruded into what was once downtown Pine Flat.

MAYOR OF PINE FLAT

Mining camps attracted folks who were different. They were often adventurous and colorful and sometimes reclusive, and more than a few sported checkered pasts. David Day was one of the original entrepreneurs in Pine Flat, opening a "fruit store" in the town in 1875. No one, not even Day's close friends, seemed to know much about where he came from or how he got to Pine Flat. Folks said he was "reticent" to talk of his past. People did, however, agree that David Day was a unique individual, and unlike most everyone else, he did not flee Pine Flat when the mines were abandoned.

In 1880, Robert Louis Stevenson found Pine Flat too remote with a multitude of inconveniences and therefore unsuited as a rent-free refuge. On the other hand, David Day and his wife settled into the vacant Reeves Hotel soon after most everyone else left town. Pine Flat remained a stop on the stage route to the Geysers, and the Days scraped out a living running a business variously described as a "station" and a "curiosity shop." They sold "canes, curios, arrowheads, milk, spruce beer and eggs" to the tourists who traveled through the Flat and to the few locals who lingered in the area.

In the summer of 1880, the Thompson Hotel caught fire. Granville Thompson and family were using the hotel as a home but were away when the fire broke out. Day was one of only several men in the town and could not stop the fire, which subsequently burned the Reeves Hotel where the Days were living, as well as a number of other buildings. Day and his wife moved into one of the surviving structures and continued their unconventional lifestyle.

In September 1889, the *Healdsburg Enterprise* carried the following story regarding a second fire:

> *Mr. Day was awakened from his sleep by the crackling of flames in the front portion of his dwelling house, at Pine Flat. He aroused his wife, telling her that the house was on fire, the roof then being almost ready to fall in. They immediately hastened from the burning building clad only in their night clothes. Mrs. Day, thinking of some valuable papers, rushed back into the house, but was caught by the devouring flames just as she reached the door and fell outside, where she was shortly afterward found with the papers clutched in her hand. Breathing the hot flames which enveloped her evidently caused her death almost instantly.*

The second fire destroyed most of the town and left David Day living in a small shack, alone and penniless. Saddened yet undaunted, Day continued to provide unique goods and services. Known for years by locals and tourists as the mayor of Pine Flat, Day was the "news center of this district." Although there was no post office in the area at the time, if anyone wanted a parcel or letter delivered to a friend, familiar words were "Leave it at old man Day's and it will be sure to reach him."

A true character, David Day was a conspicuous resident of Pine Flat for twenty-five years, from its heyday as a lively mining town to its life as a quiet, abandoned curiosity on the way to the Geysers. In May 1897, Day was stricken with paralysis and passed away in Pine Flat. With his death, a news article reported that it was "as if some old landmark has been obliterated."[249]

Chapter 8

SEARCHING FOR
HISTORICAL SIGNIFICANCE

A look back at the Sonoma County quicksilver rush of the 1870s and its lasting historical significance reveals that it did little to alter the course of California history. Its overall positive effect was a resounding "not much," which is not to say that for a short time it was not the major event that captivated Sonoma County. The local populace, fueled by eager journalists and a lingering frontier spirit, was energized in ways that resembled the "days of '49." Local merchants and entrepreneurs enjoyed a brief but significant business boom. Optimism swept the area, and prosperity and growth in mining appeared imminent. It was an exciting time.

However, once the mines closed, there was no lasting community in Pine Flat, and the roads that were built were either abandoned or little used. Prospect fever and the resulting rush was a short-lived aberration that came and went with lightning speed. With the mines closed and Pine Flat abandoned, most of the participants probably were not eager to discuss whatever hardships and misfortune they had endured. The previously hyper-supportive journalists were silent. A few short decades after the rush, it was safe to say that most Sonoma County residents knew little of Pine Flat's flash of notoriety.

The rush did not bring hordes of new immigrants to the area. Many of the prospectors were locals who temporarily dropped their plows and headed to the mountains searching for instant wealth. Afterward, they quietly returned to their agrarian endeavors. Transient Anglo and Mexican workers quickly left the area when employment opportunities vanished. Most of those who opened businesses in Pine Flat were from Calistoga and Healdsburg and also returned to their previous hometowns once the fury subsided. The Chinese had come to California for the construction of the railroads and in 1873 were living and searching for work in San Francisco. They went to the Cinnabar Mining District as a temporary labor force, and most returned to San Francisco once it ended.

Unreported by contemporary journalists and often ignored by historians, the quicksilver fever nonetheless exemplified the Manifest Destiny attitudes of Sonoma County Anglos and the resulting treatment of minorities. The frenzy erased any hope the Wappo had for retaining their Mayacamas homeland. Mexican miners, while valued workers, were referred to in condescending terms, were forced to live in segregated housing and were never part of the social fabric of Pine Flat. The Chinese experienced blatant prejudice, were exposed to toxic trauma and were discarded after their labor was not needed. They fled the area, eventual victims of the 1882 Chinese Exclusion Act.

Economically, the rush had little lasting effect on the area. Initially, it stimulated enthusiasm, but few made money on quicksilver. Some financiers from out of the area did invest in the mines, but their capital had little effect on the local economy. The small quantity of quicksilver produced did virtually nothing to influence the worldwide production or price of quicksilver. Most of the entrepreneurs who invested in supporting businesses abandoned their efforts by 1875. One might conclude that Pine Flat and its mines provided needed employment during the recession of the 1870s, and for a short time, that may have been accurate.

One lasting effect of the rush seems to have been that it illustrated and perhaps advanced the civility and emerging law related to mining claims. During the earlier gold rush of '49, there were few formal rules regulating ownership of mine claims, and disputes were often settled by spontaneous and sometimes deadly means. The disputes among the 1860s claimants in the Cinnabar Mining District and those claimants of the 1870s were settled quite differently in the courts of Sonoma County. While potentially violent arguments and one "shooting scrape" between two mining superintendents were reported, most of the action was at the spectacular trials in the county courthouse in Santa Rosa.

It is safe to assume that environmental issues created by the rush are what had the most lasting effect on the area. The Wappo had lived in Pine Flat for two thousand years prior to the Anglo invasion and preserved a balance of people, animals and plants. However, this balance was never a concern to quicksilver miners. As with most nineteenth-century mining ventures, miners had little, if any, need to balance their actions with the natural environment. The nearby mountains were stripped of timber for use in constructing tunnels and also as firewood for the retort furnaces to reduce the ore. In 1873, game was plentiful in the area with deer, elk, grizzly and brown bears. By 1875, the elk and bears had disappeared, having been killed for food by the hungry miners. These species have never returned in any significant numbers. It was said by local ranchers that it was decades before deer returned to their original numbers.

The 1870s miners reported easily catching salmon and hundreds of trout in local creeks. Mine tailings containing toxic mercury and other debris were routinely dumped into the streams, and this undoubtedly contributed to a diminished fish population in the Russian River watershed. A news report in 1875 noted, "Water from sluice pipe of the Yellow Jacket [Mine] made Mayacamas [Creek] very muddy," and, "It is said by the settlers along the creek that it destroys fish and the settlers are displeased."[250] This muddy creek flow was undoubtedly laced with toxic mercury. As recently as 2009, a government study detailed the damage and high levels of mercury detected in the Little Sulphur Creek watershed. However, given the short duration and limited success of 1870s mercury mining in Sonoma County, only part of the environmental damage can be attributed to the 1873–75 rush.

Consequently, the Sonoma County quicksilver rush of 1873–75 remains a little-known historical curiosity. It was fueled by a brief economic quirk in the quicksilver market. The rush stirred the hysteria, optimism and disappointment emblematic of countless other mining booms of the nineteenth century. However, its short duration and lack of financial success precluded it from having a significant historical legacy. For a short time, it aroused a restless local populace that was just beginning to abandon its frontier spirit roots and settle into the agrarian lifestyle that became the hallmark of Sonoma County. While the Wappo, if asked, would have had a radically different view, for traditional American historians, the quicksilver fever of the 1870s was an interesting and exciting diversion that is often overlooked. It displaced Native Americans and left some serious environmental concerns but did not do much else.

A Pine Flat Ghost

———◆━╳━◆———

Having researched and written, for far too long, about quicksilver mining and Pine Flat prospect fever in Sonoma County, I feel I know the people who were there. I appreciate the time in which they lived and worked and almost understand why they did what they did. Damn, I want to go back there—as a Pine Flat ghost!

What a thrill to survive a wild, white-knuckle ride into Pine Flat, bouncing painfully for three hours on a Clark Foss wagon seat. I'd hop off the stage, strut into J.L. Terry's rowdy saloon and call for a shot of Pine Flat whiskey that "kills at forty rods" while W.C. Graves is regaling all with a tale of Donner Party travails. I'd reluctantly beg off a saucy invite to a local "hurdy-gurdy" house and then warily venture into the Chinese camp on the edge of town just to see if they were smoking opium and playing that "fan-tan" gambling game. I'd hire a mount from A.A. Brown and head for the Rattlesnake Mine; hear B.F. DeNoon's wild predictions of imminent riches; peek into a dark, dank tunnel, check out a blazing hot furnace; and fondle some of the shiny liquid metal that is the local obsession. Hot, dry and dusty, I'd ride back to town and devour hard tack and venison stew at James Hickle's eatery. Tired as hell, I'd rent a room at the rickety Thompson Hotel and slumber restlessly on a rock-hard bed in a rough, wood-framed chamber. The next morning, if I got up the nerve, I'd search out Gran or Green Thompson, offer to buy them a drink and then cautiously query them about their scurrilous encounters in

Napa County. About noon, I'd pony up the $1.50 fare for a swift downhill stage ride back to the valley and my tranquil life on the farm.

How fortunate I am to have been a Pine Flat ghost who caught a short-lived, yet unforgettable, dose of Pine Flat prospect fever.

NOTES

A Character of the Time: Archibald Campbell Godwin

1. Compiled from Munro-Fraser, *History of Sonoma County*; U.S. Bureau of Indian Affairs publications; *Confederate Veterans Magazine*; *California Farmer and Journal of Useful Sciences* 4; Hutchings, *Scenes of Wonder*; *Sacramento Union* (1854–58); and *Petaluma Journal* (1855–60).

Chapter 2

2. Lord, *Comstock Mining*, 80–87.
3. Bowles, *Across the Continent*, 20.
4. *Sonoma Democrat*, October 30, 1875.
5. *Sonoma County Journal*, February 10, 1860.
6. California State Mining Bureau, *Fourteenth Report*, 342.
7. *Sonoma County Journal*, January 11, 1861.
8. Thompson, *Historical Atlas*, 17.
9. California State Mining Bureau, *Fourth Report*, 336.
10. Bradley, "Quicksilver Resources," 154.
11. *Mining and Scientific Press*, November 15, 1873.
12. Johnston in *Mercury and the Making of California* explains the workings of the "Magic Quicksilver Ring," 74–75.
13. *Mining and Scientific Press*, November 15, 1873.
14. Ibid., March 1, 1873.
15. Bancroft, *History of California*, 658.
16. *Mining and Scientific Press*, March 1, 1873.
17. Ibid., January 23, 1875.

18. *Napa Reporter*, July 5, 1873.
19. Hutchings, *Scenes of Wonder*.
20. *Russian River Flag*, October 23, 1874.
21. *Sonoma Democrat*, November 1, 1873.
22. *Mining and Scientific Press*, March 21, 1874.
23. Ibid., October 3, 1874.
24. Ibid.
25. Ibid., August 22, 1874.
26. *Russian River Flag*, February 12, 1874.
27. *Mining and Scientific Press*, August 22, 1874.
28. *Russian River Flag*, February 11, 1875.
29. *Napa Register*, April 12, 1873.
30. Ibid.
31. *Russian River Flag*, June 25, 1874.
32. *Napa Register*, November 1, 1873.
33. *Russian River Flag*, June 25, 1874.
34. Ibid.

A CHARACTER OF THE TIME: TIBURCIO PARROTT

35. Much has been written about Tiburcio Parrott, and this mini-biography reflects those writings. His unpublished biography, *Tiburcio Parrott 1840–1904*, by Jourden Myers can be viewed online. The *San Francisco Call* (1890–94) and *Daily Alta California* (1871–91) newspapers often reported his business and social exploits. Additional historical data can be found on the Spring Mountain Vineyard website (www.springmountainvineyard.com).

CHAPTER 3

36. *Russian River Flag*, November 6, 1873.
37. *Sonoma Democrat*, April 12, 1873.
38. *Napa Register*, April 12, 1873.
39. *Russian River Flag*, June 25, 1874.
40. *Sonoma Democrat*, November 1, 1873.
41. Ibid.
42. Ibid.
43. *Russian River Flag*, October 23, 1873.
44. *Napa Register*, May 24, 1873.
45. *Sonoma Democrat*, November 1, 1873.
46. Ibid.
47. *Russian River Flag*, October 23, 1873.
48. *Sonoma Democrat*, November 1, 1873.

49. *Mining and Scientific Press*, March 21, 1874.

50. Ibid., May 10, 1873.

51. *Napa Register*, February 14, 1874.

52. *Mining and Scientific Press*, April 4, 1874.

53. *Russian River Flag*, February 12, 1874.

54. *Sonoma Democrat*, March 7, 1874.

55. *Russian River Flag*, April 16, 1874.

56. *Napa Register*, May 9, 1874.

57. Ibid., June 13, 1874.

58. Ibid.

59. *Mining and Scientific Press*, June 13, 1874.

60. *St. Helena Star*, January 14, 1875.

61. *Calistoga Free Press*, March 14, 1874.

62. *Mining and Scientific Press*, March 14, 1874.

63. *Sonoma Democrat*, March 7, 1874.

64. Ibid., October 24, 1874.

65. *Napa Register*, November 16, 1873.

66. *Mining and Scientific Press*, March 21 and July 25, 1874; *Russian River Flag*, October 23, 1873.

67. *Russian River Flag*, January 15, 1874.

68. *Mining and Scientific Press*, November 21, 1874.

69. *Russian River Flag*, June 25, 1874.

70. Ibid., September 17, 1874.

71. *Mining and Scientific Press*, July 25, 1874.

72. Ibid., May 23, 1874.

73. Ibid.

74. *Russian River Flag*, April 16, 1874.

75. Goss, *Life and Death of a Quicksilver Mine*.

76. *Russian River Flag*, April 23, 1874.

77. *Napa Register*, September 12, 1874.

78. *Russian River Flag*, June 25, 1874.

79. Ibid., July 9, 1874.

80. Ibid., July 2, 1874.

81. Ibid.

82. Ibid.

83. Bradley, "Quicksilver Resources," 182.

84. *Russian River Flag*, June 25, 1874.

85. *Mining and Scientific Press*, July 25, 1874.

86. Ibid.

87. California State Mining Bureau, *Eleventh Report*, 132.

88. *Mining and Scientific Press*, July 25, 1874.

89. *Russian River Flag*, June 25, 1874.

A Character of the Time: Cum Yuk

90. Compiled from *Russian River Flag* (1873–75); *Healdsburg Enterprise* (1876); Lydon, *China Gold*; and Goss, *Life and Death of a Quicksilver Mine*.

Chapter 4

91. *Sonoma Democrat*, November 8, 1873.
92. Ibid.
93. *Daily Alta California*, July 6, 1873.
94. *Russian River Flag*, October 23, 1873.
95. Ibid., November 13, 1873.
96. Ibid.
97. Ibid.
98. Ibid., November 27, 1873.
99. Ibid., November 18, 1873.
100. Ibid.
101. Ibid., June 26, 1874.
102. Ibid., March 5, 1874.
103. Ibid., February 12, 1874.
104. *Napa Register*, January 24, 1874.
105. *Napa Reporter*, February 14, 1874.
106. *Russian River Flag*, January 15, 1874.
107. Ibid., March 5, 1874.
108. Ibid., January 1, 1874.
109. Ibid., October 23, 1873.
110. Ibid., February 12, 1874.
111. Ibid., March 12, 1874.
112. *Mining and Scientific Press*, March 21, 1874.
113. *Russian River Flag*, April 9, 1874.
114. Ibid., May 28, 1874.
115. Ibid., June 4, 1874; June 11, 1874; June 18, 1874.
116. Ibid., July 16, 1874.
117. *Napa Reporter*, February 14, 1874.
118. *Napa Register*, May 16, 1874.
119. *Russian River Flag*, May 21, 1874.
120. Ibid., May 28, 1874.
121. *Sonoma Democrat*, November 8, 1873.
122. *Russian River Flag*, October 1, 1874.
123. Ibid., September 24, 1874.
124. *Sonoma Democrat*, October 10, 1874.
125. Ibid., November 28, 1874.

126. Ibid., December 12, 1874.
127. Ibid., February 6, 1875.
128. Ibid., February 20, 1875; February 26, 1875; March 6, 1875.
129. Thompson, *Historical Atlas*, 17.
130. *Santa Rosa Times*, March 4, 1875.
131. *St. Helena Star*, March 11, 1875.

A CHARACTER OF THE TIME: OLD CHIEFTAN

132. Much has been written about Clark Foss, from numerous contemporary journalists to Robert Louis Stevenson in *The Silverado Squatters*. This account is taken from those sources.

CHAPTER 5

133. *Russian River Flag*, July 9, 1874.
134. Ibid., June 25, 1874; July 2, 1874; October 29, 1874; *Sonoma Democrat*, May 22, 1874.
135. *Russian River Flag*, July 9, 1874.
136. *Sonoma County Journal*, June 15, 1860.
137. *Russian River Flag*, October 23, 1873.
138. Ibid., July 9, 1874.
139. *Calistoga Free Press*, July 9, 1874.
140. *Russian River Flag*, July 9, 1874.
141. *Calistoga Free Press*, October 31, 1874.
142. *Russian River Flag*, July 9, 1874.
143. *Napa Register*, November 1, 1873.
144. *Russian River Flag*, November 27, 1873.
145. Ibid., February 12, 1874.
146. Ibid., March 26, 1874.
147. *Napa Reporter*, May 16, 1874.
148. *Russian River Flag*, April 30, 1874.
149. *Napa Reporter*, May 23, 1874.
150. Ibid., May 16, 1874.
151. Ibid.
152. *Russian River Flag*, October 1, 1874.
153. *Sonoma Democrat*, October 17, 1874.
154. *St. Helena Star*, October 23, 1874.
155. *Calistoga Free Press*, June 18, 1874.
156. *Russian River Flag*, April 2, 1874.
157. Ibid., May 14, 1874.

158. Ibid., April 1, 1875; *Calistoga Free Press*, May 15, 1875.
159. *Russian River Flag*, April 1, 1875; June 24, 1875; *Mining and Scientific Press*, June 6, 1874; August 7, 1875.
160. *Calistoga Free Press*, June 6, 1874; August 7, 1875.
161. *Russian River Flag*, October 1, 1874; April 1, 1875; *Sonoma Democrat*, October 17, 1874.
162. *Russian River Flag*, June 25, 1874; April 1, 1875; *Calistoga Free Press*, April 3, 1875.
163. *Sonoma Democrat*, October 17, 1874.
164. *Russian River Flag*, December 3, 1874.
165. Ibid., November 4, 1875.
166. Ibid., February 25, 1875.
167. *St. Helena Star*, April 8, 1875.

A CHARACTER OF THE TIME: WILLIAM C. GRAVES

168. Compiled from Findagrave.com; Kristin Johnson, *New Light on the Donner Party* website; J. D. Larimore, Graves family genealogist (Find-a-grave.com); *Russian River Flag* (1874–75); and *Santa Rosa Republican* (March 9, 1907).

CHAPTER 6

169. Issler, *Our Mountain Hermitage*, 32.
170. Stevenson, *Silverado Squatters*, 12.
171. *Russian River Flag*, May 14, 1874.
172. *Santa Rosa Times*, February 18, 1875.
173. *Russian River Flag*, March 25, 1875.
174. *Calistoga Free Press*, July 25, 1874.
175. Ibid.
176. *St. Helena Star*, December 31, 1874.
177. *Calistoga Free Press*, October 24, 1874.
178. *Russian River Flag*, March 18, 1875.
179. *Sonoma Democrat*, November 28, 1874.
180. *Russian River Flag*, July 2, 1874.
181. Ibid., February 12, 1874.
182. *Sonoma Democrat*, January 2, 1875.
183. *Calistoga Free Press*, January 2, 1875.
184. Ibid.
185. *St. Helena Star*, November 5, 1874.
186. *Santa Rosa Times*, February 18, 1875.
187. *St. Helena Star*, February 14, 1875.
188. *Napa Register*, September 12, 1874.

189. Ibid., April 25, 1874.

190. *Petaluma Weekly Argus*, October 30, 1874; *Calistoga Free Press*, October 17, 1874.

191. *Russian River Flag*, April 9, 1874.

192. *Napa Register*, March 21, 1874.

193. *Russian River Flag*, April 16, 1874.

194. *St. Helena Star*, April 8, 1875.

195. Munro-Fraser, *History of Sonoma County*, 29–30.

196. *Sonoma Democrat*, October 17, 1874.

197. *Los Angeles Herald*, September 27, 1874.

198. *Russian River Flag*, July 9, 1874.

199. Ibid., July 2, 1874.

200. Shipley, *Tales of Sonoma County*, 130.

201. *St. Helena Star*, November 26, 1874; November 31, 1874; *Calistoga Free Press*, February 20, 1875; *Russian River Flag*, July 9, 1874; May 20, 1875.

202. *Russian River Flag*, June 4, 1874.

203. Goss, *Life and Death of a Quicksilver Mine*.

204. *Sonoma Democrat*, September 18, 1875.

205. *St. Helena Star*, January 14, 1875.

206. *Calistoga Free Press*, April 17, 1875.

207. *Sonoma Democrat*, April 17, 1875.

208. Ibid., December 4, 1875.

209. *Russian River Flag*, May 16, 1872.

210. *Calistoga Free Press*, March 18, 1875.

211. *St. Helena Star*, November 5, 1874.

212. Ibid., November 12, 1874.

213. Ibid.

214. *Calistoga Free Press*, November 14, 1874.

215. *St. Helena Star*, November 12, 1874.

216. *Russian River Flag*, August 13, 1874.

217. *Calistoga Free Press*, March 18 and April 3, 1875.

218. Ibid., February 20, 1875.

219. *St. Helena Star*, October 23, 1874.

220. *Napa Register*, May 16, 1874; *Russian River Flag*, May 25, 1874.

221. *St. Helena Star*, December 31, 1874.

A CHARACTER OF THE TIME: GRAN-GREEN THOMPSON

222. Compiled from *Russian River Flag* (1873–75); *Calistoga Free Press* (1873–75); *Sonoma Democrat* (1873–75); *Napa Register* (1872–78); *Healdsburg Enterprise* (1878–80); *History of Napa and Lake Counties* (San Francisco, 1881) and U.S. census and voting records (1850–1910).

Chapter 7

223. *Russian River Flag*, March 18, 1875; April 22, 1875.
224. Ibid., May 27, 1875.
225. Ibid., December 2, 1875.
226. Ibid., November 4, 1875.
227. *Calistoga Free Press*, October 16, 1875.
228. *Sonoma Democrat*, October 30, 1875.
229. *Calistoga Free Press*, October 16, 1875.
230. California State Mining Bureau, *Fourth Report*, Table I.
231. *Healdsburg Enterprise*, August 26, 1876.
232. *Russian River Flag*, September 7, 1876.
233. Paul, *Mining Frontiers*, 80.
234. California State Mining Bureau, *Fourth Report*, Table I.
235. *Healdsburg–Pine Flat Road Assessment Rolls*, October 5, 1874.
236. Lewis, *Gypsies*, 224.
237. *Healdsburg Enterprise*, June 27, 1879.
238. *Russian River Flag*, July 8, 1880.
239. Stevenson, *Silverado Squatters*, 27.
240. Ibid.
241. Issler, *Our Mountain Hermitage*, 35.
242. Stevenson, *Silverado Squatters*, 28.
243. Issler, *Our Mountain Hermitage*, 56.
244. Issler, *Stevenson at Silverado*, 33.
245. *Healdsburg Enterprise*, July 8, 1880.
246. *Sotoyome Sun*, September 14, 1904.
247. *Geyserville Gazette*, October 27, 1916.
248. California State Mining Bureau, *Journal of Mines*, 98.

A Character of the Time: Mayor of Pine Flat

249. Compiled from *Healdsburg Enterprise* (September 18, 1889); *Healdsburg Tribune* (May 27, 1897); and *Middletown Independent* (June 4, 1897).

Chapter 8

250. *Russian River Flag*, February 18, 1875.

BIBLIOGRAPHY

BOOKS

Bancroft, Hubert Howe. *History of California*. Vol. 7. San Francisco: History Company, 1890.

Bowles, Samuel. *Across the Continent*. Springfield, MA: Samuel Bowles and Co., 1965.

Goss, Helen Rocca. *The Life and Death of a Quicksilver Mine*. Los Angeles: Historical Society of Southern California, 1958.

Hutchings, James M. *Scenes of Wonder and Curiosity in California*. San Francisco: James Hutchings & Co., 1862.

Issler, Anne Roller. *Our Mountain Hermitage*. Stanford, CA: Stanford University Press, 1950.

———. *Stevenson at Silverado*. Fairfield, CA: John Stevenson Publisher, 1996.

Johnston, Andrew Scott. *Mercury and the Making of California: Mining, Landscape, and Race, 1840–1890*. Boulder: University of Colorado Press, 2013.

Lewis, Dio. *Gypsies: Or Why We Went Gypsying in the Sierras*. Boston: Eastern Book Company, 1881.

Lord, Eliot. *Comstock Mining and Mines*. Berkeley, CA: Howell-North Press, 1959.

Lydon, Sandy. *China Gold*. Capitola, CA: Capitola Book Company, 1985.

Menefee, C.A. *Historical and Descriptive Sketch Book of Napa, Sonoma, Lake and Mendocino*. Napa City, CA: Reporter Publishing House, 1873.

Munro-Fraser, J.P. *History of Sonoma County*. San Francisco: Alley, Bowen & Co., 1880.

Palmer, Lyman L. *History of Napa and Lake Counties*. San Francisco: Slocum, Bowen & Co., 1881.

BIBLIOGRAPHY

Paul, Rodman Wilson. *Mining Frontiers of the Far West, 1848–1880*. New York: Holt, Rinehart and Winston, 1963.

Shipley, William C. *Tales of Sonoma County*. Santa Rosa, CA: Sonoma County Historical Society, 1965.

Stevenson, Robert Louis. *The Silverado Squatters*. New York: Collier Press, 1912.

Thompson, Thos. H., ed. *Historical Atlas of Sonoma County*. Oakland, CA: Thos. H. Thompson and Co., 1877.

NEWSPAPERS

California Farmer and Journal of Useful Sciences (San Francisco), September 9, 1855–October 17, 1855.

Calistoga (California) *Free Press*, April 18, 1874–October 16, 1875.

Daily Alta California (San Francisco), February 28, 1861–October 13, 1875.

Geyserville (California) *Gazette*, October 27, 1916.

Healdsburg (California) *Enterprise,* June 10, 1876–September 18, 1889.

Los Angeles (California) *Herald,* June 4, 1874–March 17, 1875.

Middletown (California) *Independent,* June 7, 1897.

Mining and Scientific Press (San Francisco), January 7, 1873–April 10, 1875.

Napa (California) *Register*, April 19, 1873–October 24, 1874.

Napa (California) *Reporter*, May 24, 1873–May 23, 1874.

Petaluma (California) *Journal*, February 2, 1856–June 4, 1856.

Petaluma (California) *Weekly Argus*, May 8, 1874–October 30, 1874.

Russian River Flag (Healdsburg, California), May 16, 1872–July 8, 1880.

Sacramento (California) *Daily Union*, May 10, 1854–September 13, 1858.

San Francisco Call, November 6, 1894.

Santa Rosa (California) *Times*, March 4, 1875–April 8, 1875.

Sonoma County Journal (Petaluma, California), February 10, 1860–January 11, 1861.

Sonoma Democrat (Santa Rosa, California), November 1, 1873–December 4, 1875.

Sotoyome Sun (Healdsburg, California), September 14, 1904.

St. Helena (California) *Star*, October 23, 1874–April 4, 1875.

OFFICIAL RECORDS

Healdsburg–Pine Flat Road Assessment Rolls. Healdsburg City Archives, Healdsburg, California, October 5, 1874.

Sonoma County. *Articles of Incorporation*. Sonoma County Courthouse, Santa Rosa, California, 1870–1875.

GOVERNMENT PUBLICATIONS

Aubury, Lewis E. "The Quicksilver Resources of California Ed." California State Mining Bureau *Bulletin No. 27*. Sacramento: California State Printing Office, 1903.

Bailey, E.H. "Quicksilver Deposits of the Western Mayacamas District, Sonoma County, California." California Division of Mines *Report 42*, San Francisco, 1946.

Bradley, Walter W. "Quicksilver Resources of California." California State Mining Bureau *Bulletin 78*. San Francisco, 1918.

California State Mining Bureau. *Eleventh Report of the State Mineralogist*. Sacramento, 1893.

———. *Fourteenth Report of the State Mineralogist*. Sacramento, 1913–14.

———. *Fourth Report of the State Mineralogist*. Sacramento, 1884.

———. *Journal of Mines and Geology*. San Francisco, 1950.

1855 Timeline of California Indian and Government Relations: Goodwin [*sic*] to Henley, January 2, 1855, Mf. RG 75, Series M234, Roll 34-316.

INTERNET SOURCES

Ancestry.com

Confederate Veteran Magazine website

Find-a-grave.com

Johnson, Kristin. "The Graves Family." New Light on the Donner Party. xmission.com/-octa/DonnerParty/Graves.htm.

LDS Family Search

INDEX

INDEX

ABOUT THE AUTHOR

Joe Pelanconi was born and raised in Sonoma County, the grandson of Italian immigrants who homesteaded in the county at the turn of the twentieth century. A career educator with a master's degree in American history, he was a secondary and college history instructor prior to a lengthy tenure as a high school principal. He has published numerous articles in educational journals and is the author of three books: *Whine & Crackers: Musings of a High School Principal*, *Vino & Biscotti: Italy to Sonoma County and Back* and *Geyserville: Fuzzy Old Snapshots*. He maintains a keen interest in local history and continues to research and publish articles in local historical publications.

Visit us at
www.historypress.net
···

This title is also available as an e-book